Workshop on Functional Materials for Cultural Heritage (Fun4Heritage 2019)

IOP Conference Series: Materials Science and Engineering
Volume 777

Matera, Italy
5 - 6 September 2019

ISBN: 978-1-7138-2665-1
ISSN: 1757-8981

Printed from e-media with permission by:

Curran Associates, Inc.
57 Morehouse Lane
Red Hook, NY 12571

Some format issues inherent in the e-media version may also appear in this print version.

Institute of Physics
Dirac House, Temple Back
Bristol BS1 6BE UK

Phone: 44 1 17 929 7481
Fax: 44 1 17 920 0979

techtracking@iop.org

Additional copies of this publication are available from:

Curran Associates, Inc.
57 Morehouse Lane
Red Hook, NY 12571 USA
Phone: 845-758-0400
Fax: 845-758-2633
Email: curran@proceedings.com
Web: www.proceedings.com

TABLE OF CONTENTS

IOP Conf. Series: Materials Science and Engineering **777** (2020) 011001 doi:10.1088/1757-899X/777/1/011001

Preface

Topic keywords

Committees

IOP Conf. Series: Materials Science and Engineering **777** (2020) 011001 doi:10.1088/1757-899X/777/1/011001

Preface

The Workshop "Functional Materials for Cultural Heritage" (Fun4Heritage, Matera, Italy, September 5-6, 2019) wishes to be an opportunity to present the state of the art in the design and synthesis of functional materials with application in the context of the restoration of cultural heritage and conservative recovery of architectural and artistic heritage.

The field of hybrid materials and functional nanocomposites is one of the most extensive multidisciplinary research areas of significant importance for the archaeological, artistic and architectural sectors, so as to find a natural place in the setting that will see Matera, in 2019, the European Capital of Culture.

Since nowadays a great emphasis is also placed on promoting technology-transfer and innovation together with spin-off/start-up company's creation, we have thought it worthwhile to dedicate a special place to SMEs, start-ups and spin-offs.

Topic keywords

- Geopolymers
- Functional Nanomaterials
- Smart Systems
- Antimicrobial Activity
- Controlled Release
- Sol-gel Technique
- Nanocomposites
- Pharmaceuticals
- Nutraceutics
- Innovative Materials

More information at the link www.ismn.cnr.it/Fun4Heritage

IOP Conf. Series: Materials Science and Engineering **777** (2020) 011001 doi:10.1088/1757-899X/777/1/011001

Committees

Scientific Committee

Germana Barone (UniCT)

http://www.dipbiogeo.unict.it/docenti/germana.barone

Sabino Aurelio Bufo (UniBAS)

http://oldwww.unibas.it/utenti/bufo/index.htm

Francesca Fontana (UniBG)

http://www00.unibg.it/struttura/strutturasmst.asp?rubrica=1&persona=311&nome=Francesca&cognome=Fontana&titolo=Prof.ssa

Pasquale Longo (UniSA)

https://docenti.unisa.it/000353/home

Candida Milone (UniME)

https://www.unime.it/it/dipartimenti/ingegneria/dipartimento/direttore

Michele Muccini (ISMN-CNR)

https://www.linkedin.com/in/michele-muccini-494b7720/?locale=en_US

IOP Conf. Series: Materials Science and Engineering **777** (2020) 011001 doi:10.1088/1757-899X/777/1/011001

Francesco Puoci (UniCAL)

http://www.unical.it/portale/portaltemplates/view/view.cfm?18749

Giovanni Salzano (UniBAS)

http://www2.unibas.it/giovannisalzano/

IOP Conf. Series: Materials Science and Engineering **777** (2020) 011001 doi:10.1088/1757-899X/777/1/011001

Organizing Committee (Chairs)

Maria Rosaria Plutino (ISMN-CNR)

mariarosaria.plutino@cnr.it

Giuseppe Rosace (UniBg)

giuseppe.rosace@unibg.it

Carmela Saturnino (UniBas)

carmela.saturnino@unibas.it

IOP Conf. Series: Materials Science and Engineering **777** (2020) 011001 doi:10.1088/1757-899X/777/1/011001

Maria Stefania Sinicropi (UniCal)

s.sinicropi@unical.it

IOP Conf. Series: Materials Science and Engineering **777** (2020) 011001 doi:10.1088/1757-899X/777/1/011001

Local Organizing Committee

M.G. Bonomo

L. Calabrone

A. Mariconda

IOP Conf. Series: Materials Science and Engineering **777** (2020) 011001 doi:10.1088/1757-899X/777/1/011001

L. Scrano

A. Vassallo

A. Caruso

IOP Conf. Series: Materials Science and Engineering **777** (2020) 011001 doi:10.1088/1757-899X/777/1/011001

D. Iacopetta

G. Sabatino

Organizing Secretariat

V. Trovato (UniBG)

A. Barbarossa, J. Ceramella (UniCAL)

M. Galletta, G. Rando (UniME)

I. Ielo (ISMN-CNR)

Sponsors acknowledgments

The Organizing Committee wishes to warmly thank the following institutions and companies for their support and contribution:

Institutional Organizers

UNIVERSITÀ DEGLI STUDI
DELLA **BASILICATA**

UNIVERSITÀ
DEGLI STUDI
DI BERGAMO

UNIVERSITÀ DELLA CALABRIA
DIPARTIMENTO DI
FARMACIA E SCIENZE
DELLA SALUTE
E DELLA NUTRIZIONE

CNR ISMN
ISTITUTO PER LO STUDIO DI MATERIALI NANOSTRUTTURATI

STVDIORVM VNIVERSITAS MESSANAE

UNIVERSITÀ DEGLI STUDI
DI SALERNO

Under the patronage of Matera municipality:

and of:

IOP Conf. Series: Materials Science and Engineering **777** (2020) 011003 doi:10.1088/1757-899X/777/1/011003

Co-Organizers

FONDAZIONE ABACUS
Istituto Europeo
di Ricerche Studi e Formazione

SMART CITIES AND COMMUNITIES
AND SOCIAL INNOVATION:
CULTURAL HERITAGE

MiUr

IOP Conf. Series: Materials Science and Engineering **777** (2020) 011003 doi:10.1088/1757-899X/777/1/011003

Supporting Sponsors

Friends and Donors

IOP Conf. Series: Materials Science and Engineering **777** (2020) 011004 doi:10.1088/1757-899X/777/1/011004

Peer review statement

All papers published in this volume of *IOP Conference Series: Materials Science and Engineering* have been peer reviewed through processes administered by the proceedings Editors. Reviews were conducted by expert referees to the professional and scientific standards expected of a proceedings journal published by IOP Publishing.

IOP Conf. Series: Materials Science and Engineering **777** (2020) 012001 doi:10.1088/1757-899X/777/1/012001

Geopolymer production for conservation-restoration using Sicilian raw materials: feasibility studies

G Barone[1*], M C Caggiani[1], A Coccato[1], C Finocchiaro[1], M Fugazzotto[1,2], G Lanzafame[1], R Occhipinti[1], A Stroscio[1] and P Mazzoleni[1]

[1] Department of Biological, Geological and Environmental Sciences, University of Catania, Corso Italia 57, Catania, Italy
[2] Department of Humanities, University of Catania, Italy, Piazza Dante 32, Catania, Italy

*E-mail: gbarone@unict.it

Abstract. In the last decades, particular attention has been aimed to innovative materials class with high-tech features to apply in conservation-restoration of Cultural Heritage, ensuring both aesthetic compatibility and adequate properties with regards to the original rock substrate. AGM for CuHe project aims to valorise Sicilian raw materials, natural and industrial by-products, for the development of alkali-activated materials, as geopolymers. These materials are produced by the activation of an aluminosilicate precursors in alkaline environment. Volcanic raw materials (volcanic ash, volcanic paleo-soils and pumice) and ceramic industrial waste were mixed with an alkaline solution and cured at room temperature. In some cases, small quantities of metakaolin were added to the mixture to increase the reactivity of the precursors. In order to obtain the geopolymeric reaction, the precursors need to be fully characterized. A similar approach is also applied to the synthesized geopolymers in order to understand its behaviour in relation to the original building materials. Therefore, a multi-analytical approach is proposed: chemical and mineralogical studies (XRF and XRD), combined with spectroscopic (FT-IR) and morphological (SEM and μCT) analyses. These preliminary results confirm the feasibility of this eco-friendly class of materials using the proposed local raw materials.

1. Introduction

In the framework of the Italian National Research Program "Advanced Green Materials for Cultural Heritage", local raw materials, as well as industrial wastes are being used as precursors for geopolymers (GP).

These materials have been recently proposed in the field of conservation-restoration interventions on built heritage in substitution to traditional mortars and to Portland cement [1–4]. Historical buildings in Sicily are characterized by brick, as well as stone structures, showing black (vulcanites) and whitish (Hyblean calcarenites) colors respectively. Moreover, mortars and plasters made using the local volcanic materials show the same hues and, in addition, a typical brick-red color which is due to the use of oxidized paleo-soils, locally named *ghiara*, as aggregates. The project aims at developing advanced, smart and high-tech products for use in high-seismic risk areas, by using the same local raw materials used in the first place for buildings and monuments. Geopolymers are versatile materials that are chemically and mineralogically compatible with traditional stone substrates. Moreover, their properties can be tailored according to the needs (blocks, mortars, plasters, etc.), in terms of viscosity, setting time, porosity, mechanical strength [1–4].

These products represent an innovative approach in terms of recycling and sustainability, since their production is carried at low temperature, involves the re-use of waste material from human activity and produces low emission of CO_2 and energy consumption in comparison with traditional Portland cements [3,5]. Moreover, the use of local raw materials reduces costs and CO_2 emissions linked to transportation. Finally, the use of colored precursors offers the possibility to obtain conservation-restoration materials that are visually similar to the original materials, yet recognizable.

Geopolymers are obtained when low-calcium reactive aluminosilicates are mixed with alkaline activators, for example sodium hydroxide and sodium silicate solutions [6]. When the silicon and aluminum contents are appropriately balanced with alkali cations and water [7,8], a geopolymeric gel will form at room temperature, through the different stages of geopolymerization reaction: dissolution of the reactive precursor, formation of oligomers, condensation of aluminosilicate units [6,7].

A multi-analytical approach is proposed to investigate the composition of the starting materials, to assess their reactivity to alkaline environment, and to optimize the formulation design. The obtained geopolymeric formulations will be then produced at a larger scale, for use in test sites. In the framework of the AGM for CuHe project, geopolymer production aims at reducing the carbon footprint and environmental impact of industrial production of conservation-restoration products, and improve the general sustainability of the intervention. The outcomes of the project encompass the development of more sustainable conservation practices, a positive impact on the industrial know-how of Sicilian companies, as well as on the tourism industry by promoting innovative fruition of cultural sites during restoration.

2. Raw materials

In this study, we assess the suitability of local georesources and of waste materials as geopolymer precursors: on the one hand, volcanic products, on the other, non-toxic waste from local sources. More in details, the raw materials used are:

- Volcanic ash, the finer fraction (< 2 mm) of particles erupted during explosive activity. It falls down on roads and is considered a waste material, stocked in landfills with high management costs.
- Volcanic paleo-soils, locally named "ghiara", widely used in Catania architecture in XVII and XVIII centuries as aggregate for the production of mortars and plasters.
- Pumice (Porticello, Aeolian Islands), a highly siliceous pyroclastic product with a characteristic "foamy" structure that is due to the simultaneous, and very rapid, depressurization and consequent cooling of the magma, generated by explosive eruptions.
- Industrial ceramic waste, linked to tiles manufacturing by local industries.

All these resources were sampled, cleaned and dried before dry milling (<75 μm). They all have an aluminosilicate composition and amorphous moieties, which are both necessary for the production of geopolymers. Therefore, a multi-analytical approach is used, which includes chemical, mineralogical, structural and microstructural characterization of the raw materials.

Moreover, commercial metakaolin has been added to geopolymers formulations.

Alkali activation was performed using appropriate mixtures of sodium hydroxide (8 M) and sodium silicate (*waterglass*) solutions.

2.1. Chemical characterization

The chemical composition of the precursors plays a significant role in the design of geopolymeric binders formulations. Portable X-ray fluorescence (PXRF) was used in this study to obtain a preliminary chemical composition of the raw materials. PXRF can be performed *in situ*, as an alternative to the more usual destructive methods based on X-rays, without any sample preparation [9]. Moreover, low costs and short analysis times are additional advantages. In detail, for this study, analyses were performed in air with a Bruker Tracer IV-SD system equipped with an Rh target X-ray tube with Pd slits and a Silicon Drif Detector (SDD). 60 s live time accumulations were collected.

Qualitative data were obtained by means of the S1PXRF software, while the abundances of the major elements by means of PyMCA software [10], which is based on fundamental parameters, considering the Tracer IV-SD specification and the measurement setup. Table 1 shows the average composition for all the raw materials.

Table 1. Chemical composition (wt%) of all the aluminosilicates precursors, from PXRF analyses; Na_2O and MgO contents are not reported due to their high analytical error.

	Al	Si	P	K	Ca	Ti	Mn	Fe
Volcanic ash	14.71	54.86	0.45	2.04	12.44	1.47	0.23	13.80
Ghiara	27.60	50.52	0.63	1.67	6.77	1.46	0.21	11.15
Ceramic waste	21.44	61.67	0.21	2.22	9.08	0.74	0.11	4.54
Pumice	14.59	75.46	0.11	4.97	2.70	0.43	0.11	1.62
Metakaolin	34.03	58.56	0.20	0.69	2.08	1.88	0.00	2.57

2.2. Mineralogical characterization

X-ray diffraction (XRD) analyses were performed on the powdered raw materials and relative geopolymers through Rietveld refinement in order to quantify the crystalline and amorphous contents (Figure 1). In detail, for this study, a Siemens D5000 diffractometer, with Cu-Kα radiation and Ni filter, in the 2θ range 3-70°, using a step size of 0.02°, a counting time of 5s for step, divergence and anti-scatter slits of 1° and receiving slit of 0.2 mm was used. The quantitative data were obtained with Rietveld method using GSASII software [11]. The amorphous and crystalline phases abundance was calculated by means of internal corundum standard addition [12].

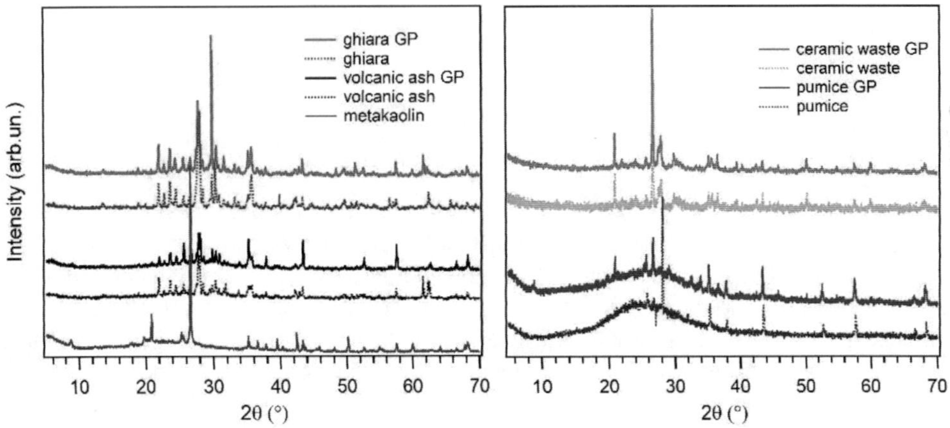

Figure 1. X-ray diffractograms of raw materials (dotted lines) and relative geopolymers (full lines).

2.3. Structural characterization

Attenuated Total Reflection Fourier-Transform Infrared Spectroscopy (FT-IR-ATR) was employed in order to exploit its high performance in the identification of chemical bonds in the structure of materials with rapid measurements.

It was performed on the powders of raw materials, used as precursors for the geopolymerization process using a Thermo Fisher Scientific 380 Infrared spectrometer (Nicolet), the spectra were collected at room temperature in the 400-4000 cm^{-1} range, with 4 cm^{-1} resolution. The ATR spectra of all the considered raw materials (Figure 2a-e) show the typical aluminosilicate structure with the broad band between 800 and 1300 cm^{-1} enveloping different T-O-T (T=Si or Al) contributions of amorphous and crystalline phases [6,13,14].

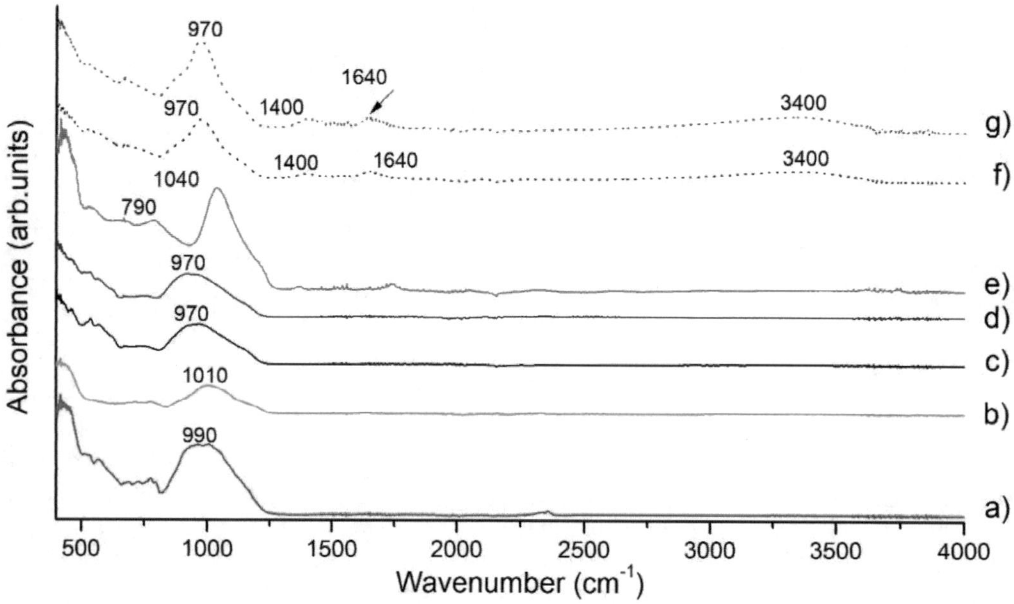

Figure 2. FTIR-ATR spectra of powdered raw materials (a: ceramic waste; b: pumice; c: ghiara; d volcanic ash; e: metakaolin) and of geopolymers synthesized with ghiara (f) or volcanic ash (g) with 20 wt% metakaolin addition.

2.4. Microstructural characterization

In this study, a Field Emission Scanning Electron Microscope (SEM) (Zeiss FEG-SEM Supra 25 Microscope) was employed to investigate samples texture from micrometric to nanometric scale. Images were collected by InLens mode using a working distance of 3/4 mm with an acceleration voltage of 20 kV and 30 kV. Analyses of the morphological features were performed on raw materials as well as on fracture surfaces of the geopolymers, covered using a gold sputter coater (8 nm thickness).

The microscopic understanding of both raw materials and "geopolymeric gel" is not straightforward, since most analyses have been performed on binders synthesized from multicomponent resources such as natural rocks or industrial waste materials which contain a large number of unreacted phases in the gel [15-17].

Volcanic ash and ghiara (Figure 3a and d) show similar texture, both containing volcanic glass. The microstructure is extremely inhomogeneous and the particles are angular and semi-angular with different sizes and shapes. Voids (Figure 3a), typically present in volcanic scoria-type grains, have dimension varying from 100 to 10 μm. Smaller particles in ghiara sample (Figure 3d) show agglomerations to form coarse grains which can be attributed to the milling process.

IOP Conf. Series: Materials Science and Engineering **777** (2020) 012001 doi:10.1088/1757-899X/777/1/012001

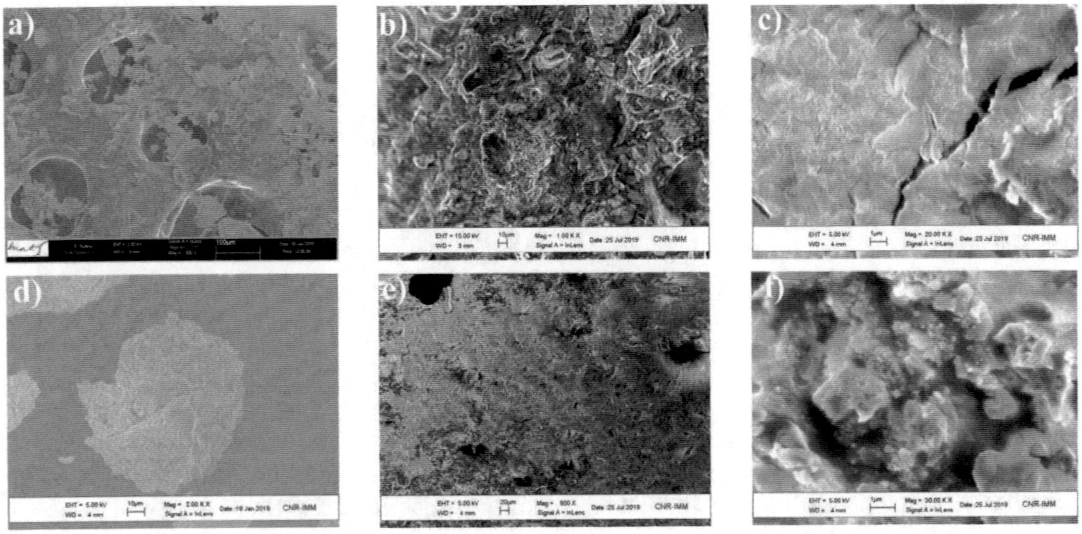

Figure 3. SEM micrographs of: a) volcanic ash raw material; b) volcanic ash-based geopolymer; c) same as b, at higher magnification; d) ghiara raw material; e) ghiara-based geopolymer; f) same as e, at higher magnification. Scale bars and magnification are shown in each image.

3. Geopolymers

The study of geopolymers needs to explore the microscopic structure of the gel at a multi-scale level, from the nanometric to the millimetric size. A multi-analytical approach is therefore needed, partially overlapping with the one used for raw materials, which encompasses XRD characterization coupled with Rietveld refinement, FT-IR analyses, SEM imaging and Synchrotron-Radiation Microtomography (SR-μCT).

As it concerns formulations design, the considered volcanic precursors are aluminium-poor (see Table 1) and they appear to be poorly reactive in alkaline environment, especially at room temperature. It has been proven that a small addition (10 to 30 wt%) of metakaolin to the mixture is sufficient to overcome this limit [15]. In this study, binary mixtures of metakaolin with volcanic ash, ghiara and pumice have been used for geopolymer formulation in order to enhance their reactivity to alkaline activation.

XRD analyses were performed on geopolymer samples after 28 days of curing in order to investigate the crystalline and the amorphous phases, and to compare them with the corresponding raw materials. The Rietveld refinement, in particular, allowed to determine the quantitative variation of these phases and consequently to establish the disappearance, decrease and change of crystalline phases and the amorphous content increase. Even though the different formulations show some variability, it is evident that the amorphous content increases in the geopolymers to the expenses of crystalline silicates (e.g. quartz, anorthite, diopside). In volcanic ash and ghiara based geopolymers, the formation of zeolites and calcite can also be appreciated. In pumice-based geopolymers, trona, a sodium carbonate hydrate, is observed as well.

The results of the ATR investigation carried out on geopolymers synthesized starting from ghiara and volcanic ash with 20 wt% of metakaolin are shown in Figure 2f-g. The broad bands around 1640 and 3400 cm^{-1} are indicative of the presence of water in the geopolymers, being respectively ascribed to H-O-H bending and to OH asymmetric stretching vibrations of molecular water [3,16]; furthermore, the band at around 1400 cm^{-1} could be attributed to the presence of carbonates. The position of the broad aluminosilicate band is shifted towards lower wavenumbers in the geopolymers (970 cm^{-1}) with respect to metakaolin (1040 cm^{-1}). However, the understanding of the formation of amorphous gel phases, and the relationship between the raw materials' (ghiara/volcanic ash, metakaolin) main ATR signal and that of the respective geopolymers requires bands deconvolution [16].

SEM observations allowed to investigate the medium to long-range ordering (tens to hundreds of nanometres). This analysis allows identifying regions of the binder that are named "geopolymeric gel" [16,18] and to study pore morphology and unreacted regions of the sample.

Differences in the morphology of corresponding geopolymer binders are hereafter shown based on high-resolution SEM images (Figure 3b, c, e, f). At this level of magnification (500-1000×), a textural feature common to both samples is the presence of micro-sized defects such as micro-voids, which may be ascribable to entrapped air and micro-cracks due to sample preparation. Both geopolymer binders contain unreacted particles bound together by geopolymeric gel. The development of an aluminosilicate gel can be observed at higher magnification (Figure 3c, f). It is characterized by a quite uniform morphology constituted by a homogeneous matrix which seems denser in volcanic ash-based geopolymer binder. The unreacted particles are loosely embedded in the matrix and they are not well distinguishable.

Geopolymers have been shown to have a microporous framework, with the characteristic pore size ranging from below 10 nm up to approximately 100 μm [19–21], which can lead to significant complexities in the pore geometry. In general, porosity can be related to the strength of a material. Control of binder's porosity can allow tailoring properties for certain applications. For this reason, SEM analyses were not intended to be used as a method to obtain statistical information on the pore size distribution [20] because they cover a small area in 2D micrographs.

A 3D information on bigger volumes should be useful to accurately determine tortuosity and connectivity. In this respect, computed microtomography has become a powerful tool for investigating the inner structure of samples at the micron scale. In the last decades, it developed among the innovative techniques employed in the characterization of geo-materials. The advantage of this technique is that it offers a rapid and non-destructive inspection of sample textures, providing parameters not obtainable by common 2D techniques, such as 3D morphology of the phases of the material and their real abundance within the investigated volume.

For the purpose of this study, feasibility tests of SR-μCT on selected samples were carried out at the SYRMEP beamline of the Elettra synchrotron laboratory (Trieste, Italy) by using high-resolution SR-μCT in phase-contrast mode [22]. A polychromatic X-ray beam in transmission geometry illuminated three millimetric-sized samples. Experiments were performed with sample-to-detector = 200 mm, collecting 1800 projections over a total scan angle of 180° with an exposure time of 2 s/projection. The effective pixel size of the detector was set at 1.952 μm^2. The 2D tomographic slices were reconstructed using the Syrmep Tomo Project (STP) house software suite [23] applying a single-distance phase-retrieval algorithm [24] to improve the consistency of the morphological analysis. The virtual 3D volumes were then segmented and analyzed using the Fijii software [25]. Analyses were performed on sub-volumes extracted from the original reconstructed 3D images. Binary images of the 3D pore morphology were obtained by manual segmentation. As a preliminary investigation, the total porosity was determined (Figure 4). Three-dimensional renderings were obtained by VGStudio Max 2.2 software (Volume Graphics, Heidelberg, Germany).

IOP Conf. Series: Materials Science and Engineering **777** (2020) 012001 doi:10.1088/1757-899X/777/1/012001

Figure 4. 3D rendering of the investigated geopolymer samples (analysed volume = 7.41 mm^3) and corresponding segmented pores phase (in yellow). The porosity φ is reported as volume fraction %.

4. Conclusions and perspectives

Locally available resources, such as pumice and ghiara, as well as waste products, such as volcanic ashes and non-toxic industrial by-products, were selected as potential aluminosilicate precursors for low-calcium alkali-activated materials, i.e. geopolymers.

The characterization of the selected raw materials allowed to optimize the design of the formulation, and to obtain geopolymeric binders with aesthetic, chemical, mineralogical, mechanical and structural compatibility with historical building materials.

The successful geopolymerization reaction is demonstrated by the increase in amorphous moieties shown by Rietveld refinement of XRD data, by the changes in the vibrational spectra, and by the observation of the gel in high-magnification electron microscopy images. Moreover, the visualization of porosity by micro-tomography is a powerful tool to understand the texture and morphological features of geopolymeric compounds.

Sicilian raw materials are suitable precursors for promising green, versatile, high-tech materials for conservation-restoration of built cultural heritage. Further studies are needed in order to assess their applicability in selected test sites, in close cooperation with local enterprises, conservators, stakeholders. Finally, particular attention will be given to positively impact the tourism industry by promoting more conscious fruition of built cultural heritage under conservation-restoration.

Acknowledgements

The research is supported by the AGM for CuHe project. PNR 2015-2020. Area di Specializzazione "CULTURAL HERITAGE". CUP E66C18000380005.

The Physics and Astronomy Department "E. Majorana" of University of Catania is acknowledged for granting access to electron microscopy facilities.

References

[1] Fernandez F, Terrana C, Enea D and Livreri P 2014 *Diagnosis, Conservation and Valorization of Cultural Heritage - Atti del quinto convegno internazionale* ed L Campanella and C Piccioli (Ariccia: Aracne editrice int.le S.r.l.) pp 266–76

[2] Miccoli L 2015 *Proc. of the 1st symp. "Knowledge Exchange for Young Scientists (KEYS)"* ed N S Msinjili and W Schmidt (Berlin: BAM Federal Institute for Materials Research and Testing) pp 519–24

[3] Clausi M, Tarantino S C, Magnani L L, Riccardi M P, Tedeschi C and Zema M 2016 *Appl. Clay Sci.* **132–133** 589–99

[4] Soragni E 2017 *Innovative Geopolymers Based on Metakaolin: Synthesis and Applications* [Dissertation thesis], Alma Mater Studiorum Università di Bologna. Dottorato di ricerca in Chimica, XIX Ciclo. DOI 10.6092/unibo/amsdottorato/7856

[5] Habert G and Ouellet-Plamondon C 2016 *RILEM Tech. Lett.* **1** 17

[6] Davidovits J 1991 *J. Therm. Anal.* **37** 1633–56

[7] Kriven W M and Schneider H 2003 *Geopolymers and Geopolymer Matrix Composites*
[8] Duxson P, Provis J L, Lukey G C, Mallicoat S W, Kriven W M and Van Deventer J S J 2005 *Colloids Surfaces A Physicochem. Eng. Asp.* **269** 47–58
[9] Shackley M 2011 *X-Ray Fluorescence Spectrometry (XRF) in Geoarchaeology* ed M S Shackley (New York, NY: Springer New York)
[10] Solé V A, Papillon E, Cotte M, Walter P and Susini J 2007 *Spectrochim. Acta - Part B At. Spectrosc.* **62** 63–8
[11] Toby B H and Von Dreele R B 2013 *J. Appl. Crystallogr.* **46** 544–9
[12] Gualtieri A F and Zanni M 1998 *Mater. Sci. Forum* **278–281** 834–9
[13] Khan M I, Azizli K A, Sufian S, Man Z, Khan A S, Ullah H and Siyal A A 2016 *Adv. Mater. Res.* **1133** 231–5
[14] Provis J and Van Deventer J eds 2009 *Geopolymers: structures, processing, properties and industrial applications* (Cambridge: Woodhead Publishing Limited)
[15] Djobo J N Y, Tchadjié L N, Tchakoute H K, Kenne B B D, Elimbi A and Njopwouo D 2014 J. *Asian Ceram. Soc.* **2** 387–98
[16] Palomo A, Grutzeck M W and Blanco M T 1999 *Cem. Concr. Res.* **29** 1323–9
[17] Yip C K and Van Deventer J 2003 *J. Mater. Sci.* **38** 3851–60
[18] Phair J W, Van Deventer J and Smith J D 2000 *Ind. Eng. Chem. Res.* **39** 2925–34
[19] Collins F and Sanjayan J G 2000 *Cem. Concr. Res.* **30** 1401–6
[20] Kriven W M, Bell J L and Gordon M 2008 *Mechanical Properties and Performance of Engineering Ceramics and Composites II - A Collection of Papers Presented at the 30th Int. Conf. on Advanced Ceramics and Composites* ed Lara-Curzio E, Zhu D, Kriven W M (Westerville, Ohio: The American Ceramic Society) pp 491–503
[21] Lloyd R R, Provis J L, Smeaton K J and van Deventer J 2009 *Microporous Mesoporous Mater.* **126** 32–9
[22] Cloetens P, Pateyron-Salomé M, Buffière J Y, Peix G, Baruchel J, Peyrin F and Schlenker M 1997 *J. Appl. Phys.* **81** 5878–86
[23] Brun F, Massimi L, Fratini M, Dreossi D, Billé F, Accardo A, Pugliese R and Cedola A 2017 *Adv. Struct. Chem. Imaging* **3** 1-9
[24] Paganin D, Mayo S C, Gureyev T E, Miller P R and Wilkins S W 2002 *J. Microsc.* **206** 33–40
[25] Schindelin J et al *Nat. Methods* **9** 676–82

IOP Conf. Series: Materials Science and Engineering **777** (2020) 012002 doi:10.1088/1757-899X/777/1/012002

Infrared spectroscopy analysis of montmorillonite thermal effects

M T Caccamo[1*], G Sabatino[1], A Bennardo[1] and S Magazù[1,2]

[1] Department of Mathematical and Informatics Sciences, Physical Sciences and Earth Sciences, Messina University, Viale Ferdinando Stagno D' Alcontres 31, 98166 Messina, Italy
[2] Istituto Nazionale di Alta Matematica "F. Severi" (INDAM), Gruppo Nazionale di Fisica Matematica (GNFM), Città Universitaria, Pl. Aldo Moro 2 - 00185 Roma, Italy

*E-mail: mcaccamo@unime.it

Abstract. The present work reports the analysis of the intramolecular OH stretching band obtained by Infrared spectroscopy measurements. In order to characterize the effect of montmorillonite on the properties of Lysozyme the so-called two-state model is adopted for the analysis of the intramolecular OH stretching band. This latter assumes that the OH stretching can be partitioned into two different states of intermolecular bonding. Such a result confirms that the montmorillonite leads to a stabilization of the Lysozyme structure. Moreover, the analysis of the spectra temperature dependence shows a montmorillonite-induced higher thermal restraint of the Lysozyme in respect to pure Lysozyme.

1. Introduction

In this work samples of montmorillonite [$(Na,Ca)_{0,3}(Al,Mg)_2Si_4O_{10}(OH)_2{\bullet}n(H_2O)$] and its mixture with Lysozyme were investigated as a function of temperature by means of Infrared spectroscopy technique.

The mineral montmorillonite, whose name derives from the town of Montmorillon, department of Vienne in France, is an Aluminium and Magnesium phyllosilicate [1-4].

Its crystalline structure is due to the overlapping, along to the C axis, of different small layers of pyrophyllite, among which are interposed one or more layers of water molecules, coordinated by a Sodium ion; this represents the so-called interchangeable ion, that can be easily substituted. Moreover, the distance between the reticular levels depends on the hydration level of the mineral; by increasing the number of water layers, the crystalline reticula expand according to the C axis, while for the complete dehydration it loses its attitude to expand.

In Nature it is possible to find this mineral in the form of very thin scales that often come together in microcrystalline masses, greasy to the touch, friable and with a remarkable absorbing power.

Among the chemical-physical characteristics [5-7] one can mention:

- Mostly white luminescence;
- Solubility in acids;
- Thixotropy.

As far as the applicative point of view is concerned, Table 1 shows some industrial fields of application and the reasons why montmorillonite is nowadays of fundamental importance for the preparation of different products.

Table 1. Some industrial applications of montmorillonite.

Montmorillonite industry applications
For removing toxins from cooking oil, due to its adsorbing and cation exchange properties
For purification of petrol and kerosene
For Waste-water treatment
For drilling mud, fireproof (suspended fire) due to the swelling properties of water
As a filler in the paper industry
Thanks to its good adhesion, it can be used instead of starch
In animal bedding, because it can absorb urine and faecal substances
In the ceramic sector where it is used as a plasticizer
For feed production
For paints production
For waterproof land in agriculture

In order to determine the thermal protective effects of montmorillonite on a fragile system of biophysical interest, a protein, i.e. the Lysozyme was chosen; this latter is an enzyme of 14,4 kilodaltons present in animal tissue and equipped with bactericidal activity; as such it can damage the bacterial cell wall of some bacteria (Gram +) catalyzing the hydrolysis of beta linkages 1,4 between N-acetylmuramic acid (NAM) and N-acetylglucosamine (NAG) which are the main component of peptidoglycan [8-11]. Its structure is constituted by two domains: one made by alpha-helix, the other one made by one beta antiparallel layer and by two alpha-helix [12-15]. The tridimensional configuration of the molecule is maintained by the presence of 3 disulfide, two located in the alpha-helix domain and one in the beta layer one, and in its internal part there are few polar residues [16-19].

2. Spectroscopic techniques

As far as the employed technique is concerned, that is infrared spectroscopy, it is a powerful tool to investigate the structural and dynamical properties of material systems as well as their dependence on temperature. The technique is based on the analysis of the absorption spectra [20-23]. In this work, we collected vibrational spectra by means of the Vertex 70v spectrometer (Bruker) using a Platinum diamond ATR [24-28]. All spectra were collected by using an average of 96 scans with a resolution of 4 cm^{-1} in a spectral range of 400÷4000 cm^{-1}, from a temperature of 25°C to a temperature of 60°C. The data were first processed through Bruker OPUS software and then by means of Matlab.

3. Spectral Distance

To characterize the temperature sensitivity of the investigated samples, the Spectral Distance approach has been employed; this latter is essentially based on the following equation:

$$SD = \sqrt{\sum_{v} [I(\omega, T) - I(\omega, T_i)]^2 \cdot \Delta\omega} \tag{1}$$

where $I(\omega)$ denotes the absorbance at the frequency ω, T_i is the lowest temperature, that in this case is equal to 25°C and $\Delta\omega$ represents the frequency resolution of the instrument.

4. Discussion and Results

Figure 1 reports FTIR spectra of montmorillonite in the $400<\Delta\omega< 4000cm^{-1}$ and in the temperature window from 25°C to 60°C.

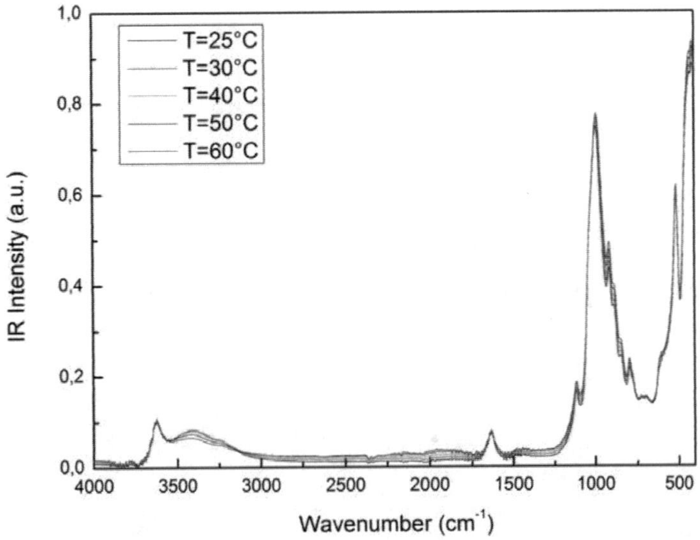

Figure 1. IR absorption intensity as a function of wavevector of montmorillonite, in the $400<\Delta\omega< 4000cm^{-1}$ from 25°C to 60°C.

Figure 2 reports FTIR spectra of Lysozyme in the $400<\Delta\omega< 4000cm^{-1}$ from 25°C to 60°C.

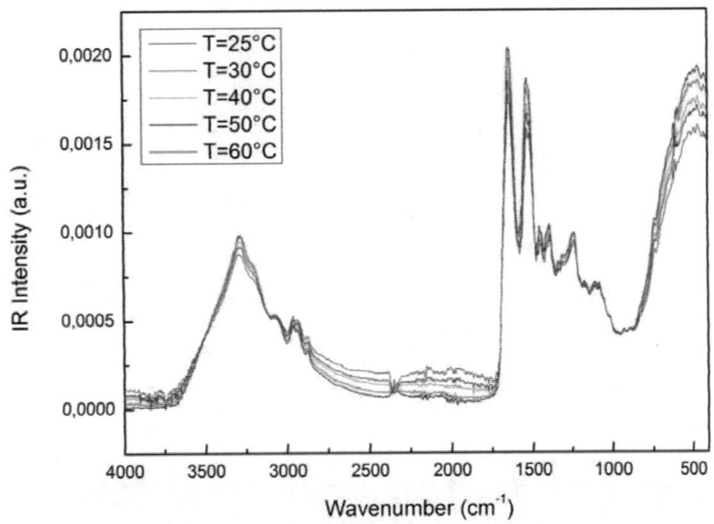

Figure 2. IR absorption intensity as a function of wavevector of Lysozyme in the $400<\Delta\omega< 4000cm^{-1}$ from 25°C to 60°C.

Finally, in Figure 3, FTIR spectra of montmorillonite in the presence of Lysozyme in the spectral range of 400-4000cm^{-1} from 25°C to 60°C is reported.

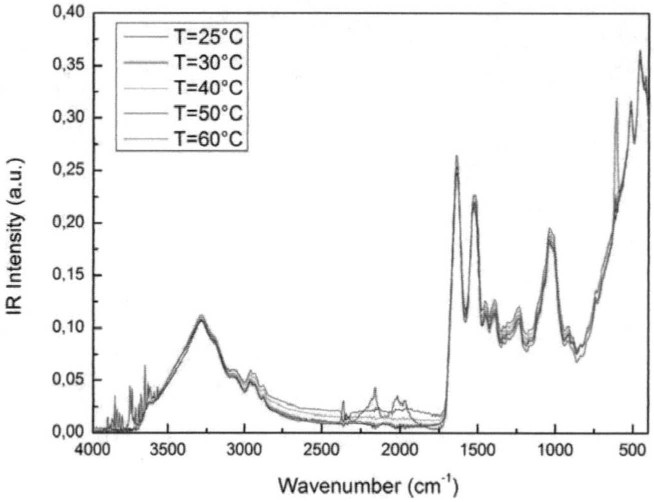

Figure 3. IR absorption intensity as a function of wavevector of montmorillonite in presence of Lysozyme in the $400<\Delta\omega<4000$cm^{-1} from 25°C to 60°C.

Concerning the spectra analysis, at first, a pre-processing data to eliminate the background has been performed. Then, in order to characterize the interaction mechanisms among montmorillonite and Lysozyme, the spectrum of montmorillonite has been subtracted from the spectrum of montmorillonite in presence of Lysozyme, so obtaining a new spectrum, i.e. the spectrum difference [29-33]. Such a procedure was performed for all the spectra at each investigated temperature. Figure 4 reports, as an example, the spectral difference for T=25°C, 30°C,40°C and 60°C.

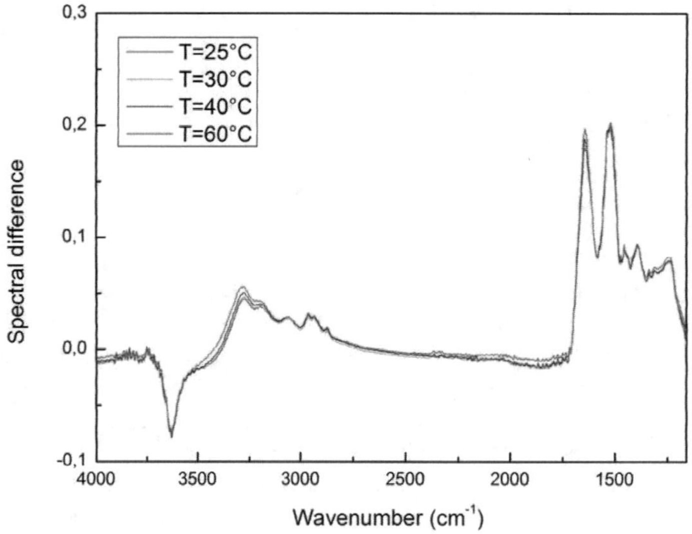

Figure 4. Spectral difference for T=25°C, 30°C,40°C and 60°C.

To compare spectra, the Spectral Distance (equation1) has been evaluated. In Figure 5, the plot of the SD versus temperature is reported both for the spectrum differences (red points) and for spectra of Lysozyme (black squares).

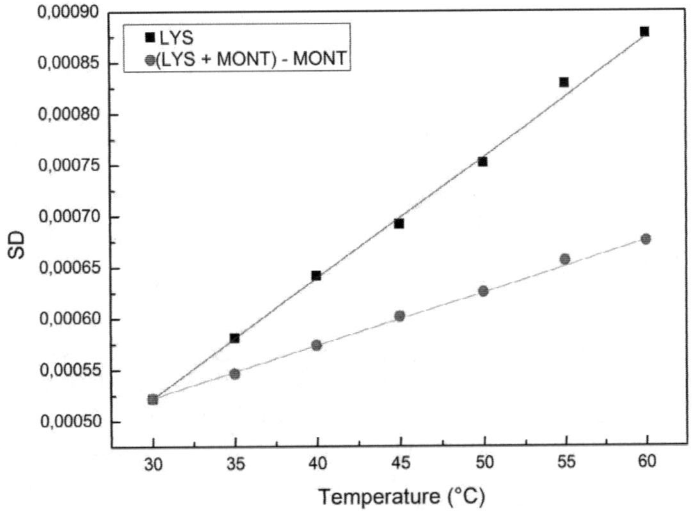

Figure 5. Spectral Distance (SD) versus temperature both for spectra difference (red points) and for spectra of Lysozyme (black squares).

As it can see, data arrange themselves along a linear behaviour, for both the investigated samples. To extract quantitative information, a linear fit has been performed:

$$y = mx + q \tag{2}$$

From this analysis, it emerges that the slope coefficient of the two systems is different; in particular, the slope value of Lysozyme in the presence of montmorillonite, $m = 5.12 \cdot 10^{-6}$, results smaller than the slope coefficient of Lysozyme, that is $m = 1.09 \cdot 10^{-5}$. Figure 5 also reports the linear fit of the two investigated systems. This result reveals that the Lysozyme in presence of montmorillonite has a smaller sensibility to temperature changes. Therefore, montmorillonite can be considered as an effective thermal bioprotector.

5. Conclusions

The present work has allowed to put into evidence the importance of clay systems as bioprotectors. A spectroscopic investigation was carried out by infrared spectroscopy on the interaction of montmorillonite with Lysozyme. It emerged that IR spectroscopy and the SD approach constitute a very effective approach for the characterization of the thermal properties of pure Lysozyme and of Lysozyme in the presence of montmorillonite. The results allowed us to see how Lysozyme in the presence of montmorillonite has a lower spectral sensitivity when the temperature changes and, hence, the role of montmorillonite as a bioprotector is thus justified.

Acknowledgments

The research is supported by the AGM for CuHe project. PNR 2015-2020. Area di Specializzazione "CULTURAL HERITAGE". CUP E66C18000380005.

References

[1] Fernández M, Alba M D and Torres Sánchez R M 2013 *Colloid Surfaces A* **423** 1

[2] Andronescu C, Alexandra Gârea S, Deleanu C, Nicolescu A and Iovu H 2013 *Appl. Clay Sci.* **86** 99
[3] Chen Q, Zhu R, Ma L, Zhou Q and He H 2017 *Appl. Clay Sci.* **135** 129
[4] Fan M, Yuan P, Zhu J, Chen T, Liu D 2009 *Journal of Magnetism and Magnetic Materials* **321** 3515
[5] Tunç S and Duman O, 2010 *Appl. Clay Sci.* **48** 414
[6] Hosseini M G, Jafari M and Najjar R 2011 *Surf. Coat. Tech.* **206** 280
[7] Komatsu L G H, Oliani W L, Lugao A B and Parra D F 2014 *Radiat. Phys. Chem.* **97**, 233
[8] Barreca D, Laganà G, Magazù S, Migliardo F, Gattuso G and Bellocco E 2014 *Int. J. Biol. Macromol.* **63** 225
[9] Öztürk N, Şenay R H, Avcıbaşı N, Akgöl S and Denizli A 2012 *Process Biochem.* **47** 816
[10] Magazù S, Migliardo F, Barreca D, Bellocco E and Laganà G 2008 *Physica B* **403** 2408
[11] Dembczynski R, Bialas W and Jankowski T 2010 *Biochem. Eng. J.* **51** 24
[12] Fenimore P W, Frauenfelder H, Magazù S, McMahon B H, Mezei F, Migliardo F, Young R D and Stroe I 2013 *Chem. Phys.* **424** 2
[13] Zhang Y, Wang Z, Zhang J, Chen C and Zhang X 2011 *Carb. Polym.,* **85** 554
[14] Magazù S, Migliardo F, Benedetto A and Vertessy B 2013 *Chem. Phys.* **424** 26
[15] Yakup Arıca M, Yılmaz M, Yalçın E and Bayramoğlu G 2004 *J. Chromatogr. B* **805** 315
[16] Magazù S, Migliardo F, Benedetto A, Mondelli C and Gonzalez M A 2011 *J Non-Cryst. Solids* **357** 664
[17] Merlini G and Bellotti V 2005 *Clin. Chim. Acta* **357** 168
[18] Ibáñez R, Almécija M C, Guadix A, Guadix E M 2007 *Sep Purif Technol* **57** 314
[19] Caccamo M T and Magazù S 2019 *AAPP* **S1**A2
[20] Caccamo M T and Magazù S 2017 *Appl. Spectr.* **71** 401
[21] Mäder K 2005 *Adv. Drug Deliver Rev.* **57** 1083
[22] Caccamo M T and Magazù S 2017 *Spectr. Letters*, **50 130**
[23] Pereira A C, Prioli L M, da Silva W J, de O. Neto G and Alvarado-Gil J J *Plant Science* **96** 203
[24] Liu K, George S C, Lu X, Gong S and Gui L 2014 *Org. Geochem.* **72** 34
[25] Gonzàlez-Jimènez J 1994 *Chem.Biol. Inter.* **91** 65
[26] Casoli A and Fornaciari S 2014 *Microchem. J.* **116** 24
[27] Bugay D E 2001 *Adv. Drug Deliver Rev* **48** 43
[28] Arnosti C, Keith S C and Blough N 2000 *Marine Chem.* **71** 321
[29] Migliardo F, Caccamo M T and Magazù S 2014 *Food Biophys* **9** 99
[30] Caccamo M T and Magazù S 2016 *Vib. Spectr.* **85** 222
[31] Caccamo M T and Magazù S 2017 *Polym. Testing* **62** 311
[32] Caccamo M T, Zammuto V, Gugliandolo C, Madeleine-Perdrillat C, Spanò A and Magazù S 2018 *Int. J. Biol. Macromol.* **114** 649
[33] Magazù S, Migliardo F and Caccamo MT 2012 *J. Phys. Chem. B* **116** 9417

IOP Conf. Series: Materials Science and Engineering **777** (2020) 012003 doi:10.1088/1757-899X/777/1/012003

Design, synthesis and characterization of hybrid coatings suitable for geopolymeric-based supports for the restoration of cultural heritage

I Ielo[1,*], M Galletta[2], G Rando[2], S Sfameni[3], P Cardiano[2], G Sabatino[4], D Drommi[2], G Rosace[5] and M R Plutino[1]

[1] Institute for the Study of Nanostructured Materials, ISMN–CNR, O.U. Palermo, c/o Department of ChiBioFarAm, University of Messina, Viale F. Stagno d'Alcontres 31, Vill. S. Agata, 98166 Messina (Me), Italy.
[2] Department of ChiBioFarAm, University of Messina, Viale F. Stagno d'Alcontres 31, Vill. S. Agata, 98166 Messina, Italy.
[3] Department of Engineering, University of Messina, C.da Di Dio, Vill. S. Agata, 98166 Messina, Italy.
[4] Department of Mathematical and Computer Sciences, Physical Sciences and Earth Sciences, University of Messina, Viale F. Stagno d'Alcontres 31, Vill. S. Agata, 98166 Messina, Italy.
[5] Department of Engineering and Applied Sciences, University of Bergamo, Viale Marconi 5, 24044 Dalmine (BG), Italy.

*Email: ileana.ielo@ismn.cnr.it

Abstract. Geopolymers are inorganic materials that form long-range, covalently bonded, networks; they are materials similar to ceramics, whose components are mainly aluminium and silicates. Any inorganic source of silica and alumina, dissolved in the alkaline solution, acts as a precursor of geopolymers. The type of clays most commonly used in the synthesis of geopolymers is based on kaolinite. The present work concerns the modification of the geopolymer at the chemical and nanostructural level, through the design and development of hybrid coating sol-gel formulations to improve characteristics such as mechanical strength, chemical inertness, heat resistance and porosity for possible use in the restoration and conservation of cultural heritage. The methods of functionalization of the geopolymers here shown are: (i) the treatment of the geopolymeric surface, post-drying, through the application of the sol-gel directly on the monolith, for a modification of the surface properties; (ii) the pretreatment of the geopolymeric mixture, through the addition of the alkoxysilane which can induce a homogeneous consolidation of the molecular structure, implementing the properties of the entire monolith. The employed alkoxysilanes differ according to the length of the hydrocarbon chain and the presence of halogen atoms. Hydrophobicity tests were performed on these formulations.

1. Introduction

The term "geopolymer" describes a wide variety of natural or synthetic composite materials showing a polymeric structure [1-3]. They are materials similar to ceramics, whose components are mainly aluminium-natural silicates (see figure 1) [4-6], although this class of compounds can easily be extended to all inorganic polymers present in nature, such as phosphates (apatites), borates, vanadates and arsenates [7-13]. The geopolymerization takes place in three phases: dissolution of the aluminosilicate source, polycondensation of aluminosilicate oligomers and finally precipitation of geopolymeric particles, which form a gel. Geopolymers have many advantages over traditional organic polymers, deriving their main characteristics from the starting mineral components and improving their chemical-physical and mechanical properties [14-21]. The sol-gel synthesis is an eco-friendly approach to functionalize geopolymers, without high-temperature treatments, perfectly in accordance with the principles of circular economy and green chemistry [22-25]. The alkoxysilane agents may be chosen for the implementation of specific properties of the geopolymer materials that allow the production of various types of coatings, such as protective coating, reflective or anti-reflective coatings, refractory linings, coatings with controlled porosity; all these properties may result determining in the restoration of cultural heritage [26-27].

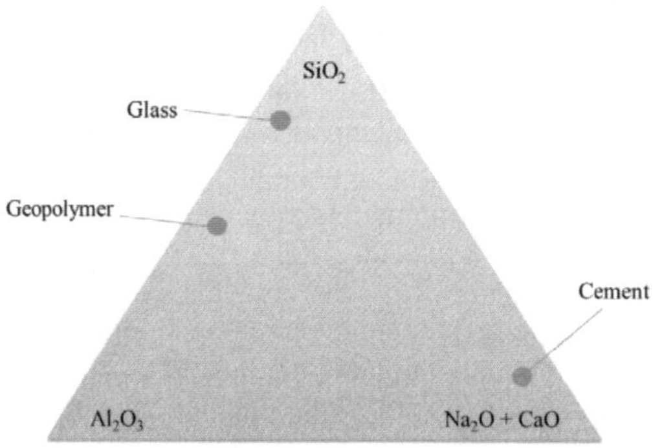

Figure 1. Geopolymer composition scheme.

2. Experimental section

2.1 Materials and Methods

Any inorganic source of silica and alumina, dissolved in the alkaline solution, acts as a precursor of geopolymers and therefore lends itself to geopolymerization [20-21].

Raw materials used in the geopolymerization process are: Kaolin, Diatomite, Montmorillonite, Volcanic Ash and Blast furnace slag. Each of the materials listed above is alkaline activated, using as alkaline activators NaOH 7M (28% w/w), 9M (36% w/w), and KOH 7M (39.2% w/w), 9M (50,4% w/w) and geopolymerized, keeping the temperature below 100°C. NaOH, KOH, *Tetraethyl orthosilicate* (TEOS, Figure 2), *(3-Aminopropyl) triethoxysilane* (APTES), (3,3,3-trifluoro-propyl) trimethoxy silane (F3) and *1H,1H-2H,2H-perfluorooctyl triethoxy silane* (F13) were purchased at the highest purity level and used as received by STREM Chemicals, inc. without any further purification.

X-ray diffraction analysis (XRDP) was performed by a BRUKER D8 ADVANCE diffractometer with radiation $K\alpha$ of Cu with Bragg-Brentano theta-theta geometry, equipped with a SiLi detector at solid state "Sol-X". The acquisition conditions are 40 kV and 40 mA. The diffractograms were obtained with scans from 2 to 70/80 degrees 2θ, with stepsize of 0.02 degrees 2θ, with a count time of 1 second. The diffractograms were filtered by $K\alpha 2$ (Fourier filter). The positions of the peaks were compared with those of the database ICDD JCPDS.

Geopolymerization process is schematized in Figure 3 and consists of three steps:

1. Dissolution in alkaline solution, to induce the exfoliation of the tetrahedral layers of aluminates and silicates;
2. Condensation in aqueous solution with formation of a gel phase;
3. Solidification of the gel in a modified 3D structure with the intercalation of water and Na^+ and K^+ cations between the aluminosilicates layers.

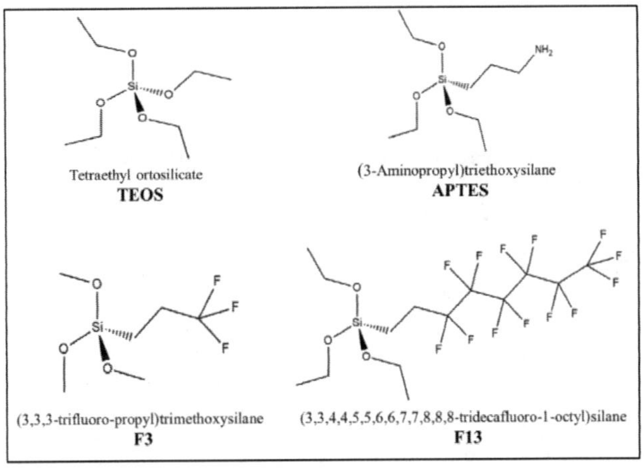

Figure 2. Sol-gel alkoxysilane precursors.

A typical procedure [28] to prepare the sol-gel solutions is as follows: TEOS (2.013g) and APTES (0.943g), together with an equimolar amount of an organic silane (F3 or F13), was dissolved in 25 mL ethanol. The resulting mixture was vigorously stirred at room temperature for 12 hours, and then ultrasonicated for 30 min to get a homogeneous suspension, and then used as coating or in addition to the geopolymeric mixture.

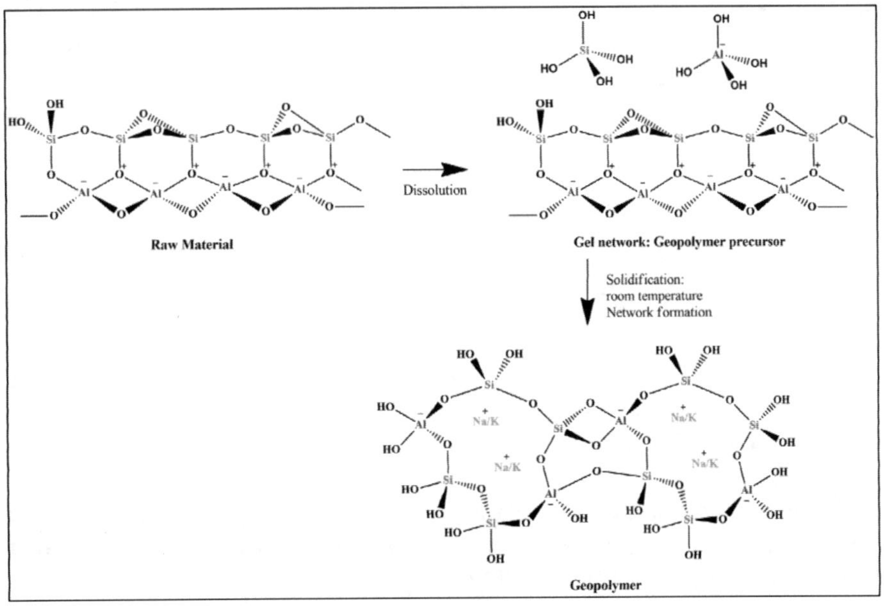

Figure 3. The chemistry of geopolymerization process.

3. Results and discussion

3.1 Sol-gel synthesis and geopolymer functionalization

As shown before [29-35] for similar functional silica based coatings, the employed sol–gel mixtures were made from *Tetraethyl orthosilicate* (TEOS) and *(3-Aminopropyl) triethoxysilane* (APTES) precursors undergoing hydrolysis–condensation reactions in several steps in combination with two alkoxysilanes, namely (3,3,3-trifluoro-propyl) trimethoxy silane (F3) and *1H,1H-2H,2H-perfluorooctyl triethoxy silane* (F13), employed individually. Associated condensation reactions are directly related with the removal of water from the polymer matrix resulting in the formation of more stable silica 3D-networks, through drying and curing processes performed at room temperature for 24h or at T=100°C for 30' following by further 30' at T=140°C.

The most important benefit of using sol-gel method concern the formation of alkoxysilane layer or xerogel at room temperature. Since there are no decomposition products formed, one can produce thick films or composite with little stress and low energy consumption. A conventional synthesis for the functionalization of geopolymers (Montmorillonite is used as aluminosilicates source) can be performed by following two different synthetic approaches, as following: (i) the treatment of the geopolymeric surface, post-drying, through the deposition of the sol-gel directly on the monolith, producing a thin film for a modification of surface properties; (ii) the pretreatment of the geopolymeric mixture through the addition of the alkoxysilane sol-gel coatings which can induce a homogeneous consolidation of the molecular structure, implementing the properties of the entire monolith. The morphology of the geopolymer surface depends on the degree of cross-linking of its molecular structure and varies according to the quantity and type of alkoxysilane added. A green method to implement these properties consists of incorporation of coupling silane agents and organic fillers during the geopolymerization process. Additionally, the surface treatment is carried out using the above discussed alkoxysilanes mixtures, which differ according to the length of the hydrocarbon chain.

3.2 Characterization

The preliminary characterization of the clays was carried out by means of powder X-ray diffraction analysis (XRDP), thus identifying the type of clay and its chemical composition (figure 4). The peaks resulting from the diffraction spectrum on powders for clay minerals can be distinguished by intense diffraction peaks, such as: quartz, feldspars, zeolites, carbonates. Mixed-layered clays are formed with two or more types of clays that alternate in overlapping layers and are more difficult to interpret by XRDP analysis, so more analyses are generally required.

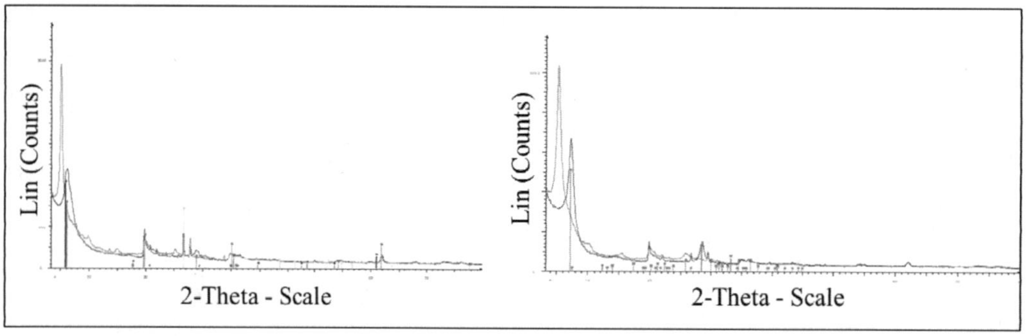

Figure 4. Powder X-ray diffraction spectra (XRDP) for two types of clays (Montmorillonite) before (blue line) and after (red line) functionalization, with TEOS_APTES/F3 formulation on left side and TEOS_APTES/F13 formulation on right side, respectively.

In both diffractograms, the first peak detected suffered a left shift in the functionalized clay, thus indicating a probable removal of the layers characterizing the mother clay and therefore an internal functionalization of the same.

3.3 Wettability tests

In the structured polymer, the alkoxysilane acts as a network inductor or as a network modifier inducing new properties to the matrix. The structural differences reflect the thermal hydrophobic properties of material determined by the addition of perfluorinated or long chain alkoxysilane species. These formulations can improve characteristics such as mechanical strength, chemical inertness, heat resistance and porosity for possible use in the restoration and conservation of cultural heritage, as well as antimicrobial activities, low adhesive properties and no biocidal effects towards bacteria. In particular, the obtained materials have a surface that shows a hydrophobic nature (see figure 5 for preliminary hydrophobicity tests). Further studies on mechanical and antibacterial properties are actually in progress.

Figure 5. Hydrophobicity tests on Montmorillonite monolith treated with the sol-gel formulations. In the image above left, surface geopolymer is treated with TEOS_APTES/F3 formulation; in the image above right, surface geopolymer is treated with TEOS_APTES/F13 formulation; in the image below, the surface is treated with TEOS only.

4. Conclusions

The aim of this work was to synthesize a new functional geopolymeric matrix based on raw materials (like montmorillonite or other clays, volcanic ashes or Pozzolan, and waste materials like fly ashes or blast furnace slags), and hybrid sol-gel to modify its properties. The employed sol-gel synthesis utilizes an eco-friendly approach to prepare and functionalize geopolymers, without high-temperature treatments. These kinds of formulations are widely used for their hydrophobic, anti-abrasive, antibacterial and antifouling action. In particular, in the first phase of this work, we tested a simple procedure for the production of a polymeric hybrid sol-gel, which was applied to the geopolymers both in the pre-treatment phase and after drying. The effectiveness of the previously mentioned properties is evaluated by wettability tests, while further studies are in progress.

Acknowledgments

The research is supported by the AGM for CuHe project. PNR 2015-2020. Area di Specializzazione "CULTURAL HERITAGE". CUP E66C18000380005. CNR and Noxorsokem Group S.r.l. are gratefully acknowledged for S. Sfameni Industrial PhD funding.

References

[1] Lee W K W and van Deventer J S J 2002 *Coll. Surf. A* **211** 115–26
[2] Plueddemann E P 1982 *Silane Coupling Agents* (New York: Springer Plenum Press) pp 111–39
[3] Dedzo G K and Detellier C 2016 *Appl. Clay Sci.* **130** 33–9
[4] Rosace G, Cardiano P, Urzì C, De Leo F, Galletta M, Ielo I and Plutino M R 2018 IOP Conf. Series: Materials Science and Engineering vol 459 012021
[5] Zarate-Reyesa L, Lopez-Pacheco C, Nieto-Camachoc A, Palaciosd E, Gómez-Vidalese V, Kaufholdf S, Uferf K, García Zepeda E and Cervini-Silvag J 2018 *J. Hazard. Mat.* **342** 625–32

[6] Cesur S, Alp B, Küçükgöksel Y, Kahraman T and Balköse D 2015 *J. Vinyl Addit. Tech.* **22** 174–82
[7] Roghanian N and Banthia N 2019 *Cem. Concr. Compos.* **100** 99–107
[8] Oyanedel-Craver V, Narkiewicz S, Genovesi R, Bradshaw A and Cardace D 2014 *J. Environ. Chem. Eng.* **2** 841–8
[9] Mejía H F G, Yohai L, Pedetta A, Seitz K H, Procaccini R A and Pellice S A 2017 *J. Colloid Interface Sci.* **508** 332–41
[10] Haile T, Nakhla G, Allouche E and Vaidya S 2010 *Corrosion Sci.* **52** 45–53
[11] Edraki M and Zaarei D 2018 *J Nanoanalysis* **5** 26-35
[12] Boyer C J, Ambrose J, Das S, Humayun A, Chappidi D, Giorno R and Mills D K 2018 *Med. Devices* **11** 123–37
[13] Noeiaghaei T, Dhami N and Mukherjee A, 2017 *Constr. Build. Mater.* **150** 880–91
[14] Abdullayev E, Sakakibara K, Okamoto K, Wei W, Ariga K and Lvov Y 2011 *ACS Appl. Mater. Interfaces* **3** 4040–6
[15] Huang J, Huang G, An C, He Y, Yao Y, Zhang P and Shen J 2018 *Environ. Pollut.* **238** 52-62
[16] He Y, Huang G, An C, Huang J, Zhang P, Chen X and Xin X 2018 *Sci. Total Environ.* **616–7** 1628–37
[17] Mejía H F G, Procaccini R A and Pellice S A 2018 *J. Solgel Sci. Technol.* **85** 529–38
[18] Roy A, Butola BS and Joshi M, 2017 *Appl. Clay Sci.* **146** 278–85
[19] England M W, Sato T, Urata C, Wang L and Hozumi A 2017 *J. Colloid Interface Sci.* **505** 566–76
[20] Robayo-Salazar R A, Mejía-Arcila J M and de Gutiérrez R M 2017 *J. Clean. Prod.* **166** 242-52
[21] Dedzo G K and Detellier C 2016 *Appl. Clay Sci.* **130** 33–9
[22] Bignozzi M C, Manzi S, Natali M E, Rickard W D A and van Riessen A 2014 *Constr. Build. Mater.* **69** 262–70
[23] Bermúdez Y H, Truffault L, Pulcinelli S H and Santilli C V 2018 *Appl. Clay Sci.* **152** 158–65
[24] Thue P S, Sophia A C, Lima E C, Wamba A G N, de Alencar W S, dos Reis G S, Rodembusch F S and Dias S L P 2018 *J. Clean. Prod.* **171** 30–44
[25] Cardiano P, Ponterio R C, Sergi S, Lo Schiavo S and Piraino P 2005 *Polymer* **46** 1857–64
[26] Davis S R, Brough A R and Atkinson A 2003 *J. Non-Cryst. Solids* **315** 197–205
[27] Castellano A, Colleoni C, Iacono G, Mezzi A, Plutino M R, Malucelli G and Rosace G 2019 *Polym. Degrad. Stab.* **162** 148–59
[28] Wang H, Ding J, Xue Y, Wang X and Lin T 2010 *J. Mater. Res.* **25** 1336–43.
[29] Rosace G, Trovato V, Colleoni C, Caldara M, Re V, Brucale M, Piperopoulos E, Mastronardo E, Milone C, De Luca G and Plutino M R 2017 *Sens. Actuators B-Chem.* **252** 428–39
[30] Rosace G, Guido E, Colleoni C, Brucale M, Piperopoulos, E, Milone C, Plutino MR, *Sens. Actuators B-Chem.* **241** 85–95
[31] Guido E, Colleoni C, De Clerck K, Plutino M R and Rosace G 2014 *Sens. Actuators B-Chem.* **203** 213–22
[32] Plutino M R, Colleoni C, Donelli I, Freddi G, Guido E, Maschi O, Mezzi A and Rosace G 2017 *J. Colloid Interface Sci.* **506** 504–17
[33] Trovato V, Colleoni C, Castellano A and Plutino M R 2018 *J. Solgel Sci. Technol.* **87** 27–40
[34] Libertino S, Plutino M R and Rosace G 2018 *2nd NanoInnovation 2017 Conf. and Exhibition* (Rome Italy: AIP Conf. Proceedings) Vol 1990 020016
[35] Castellano A, Colleoni C, Iacono G, Mezzi A, Plutino M R, Malucelli G and Rosace G 2019 *Polym. Degrad. Stab.* **162** 148–59

IOP Conf. Series: Materials Science and Engineering **777** (2020) 012004 doi:10.1088/1757-899X/777/1/012004

Silver fir characterized by micro-imaging NMR and FTIR spectroscopy

S Longo[1,2,*], S Capuani[2], C Corsaro[1] and E Fazio[1]

[1] Department of Mathematical and Computational Sciences, Physics Science and Earth Sciences (MIFT), University of Messina, Viale Ferdinando Stagno D' Alcontres 31, 98166 Messina, Italy.
[2] Nuclear Magnetic Resonance and medical physics Laboratory CNR-ISC, Department of Physics, Sapienza University of Rome, Piazzale Aldo Moro 5, 00185 Rome, Italy.

[*] E-mail: sveva.longo@unime.it

Abstract. Nuclear Magnetic Resonance micro-imaging (μMRI) approach with multiparametric measurement and FTIR spectroscopy was adopted to investigate silver fir. In particular, we investigated weakly interacting and bound water, detected by NMR and FTIR techniques, hypothesizing a water interaction with the main structures of the wood. For instance, by means of NMR analyses, we obtain information about the principal wood morphological/structural modifications, which are associated with humification processes, reflected in the changes of aromatic C-H and C-O vibration modes shown by FTIR. Thus, the combined use of NMR and FTIR allows characterizing the mobility of water with respect to lignin and cellulose features, which can serve as a benchmark for wood degradation conditions.

1. Introduction

The study of wood is recently an emerging area in materials science. Wood as a natural filler is a low-cost material with low density and high specific properties, and it is also biodegradable and non-abrasive. Wood is also characterized by a good strength to weight ratio. It is shock-resistant and has the ability to bend without fracture [1]. For all these properties, it has been largely used for the manufacturing of many historical works which nowadays require timely actions to preserve their state of beauty. One of the main wood structural disadvantages occurs when it is penetrated by moisture and water [2]. Hence, it is interesting to define the relationship between the morphological characteristics and the micro- and ultrastructure of wood tissues, as expressed by the thickness of tracheid walls, lignin content and crystallinity of cellulose as well as by its hydrophobic capacity. Most of the standard techniques used to characterize wood morphology allows only in part to clearly define the bulk and surface characteristics with its porosity and, in turn, to the hydrophobic capacity of woods [3].

In this work, micro-NMR imaging analyses were carried out on a piece of silver fir (*Abies alba*) and some of the information, obtained by NMR relaxation times T_1 and T_2 and diffusion coefficient D of water in wood and those collected by conventional FTIR method are discussed.

2. Experimental

Morphological and structural analyses were carried out on silver fir (*Abies alba*) samples totally immersed in distilled water. The silver fir was purchased by Pircher Oberland Spa, a company from

Dobbiaco, in the heart of Alto Adige (Italy), which deals with wood processing to produce cutting-edge products, also for domestic use.

2.1. NMR imaging acquisition

Magnetic resonance microimaging (μMRI) analyses on a sample of silver fir (*Abies alba*) of 2x0,5 cm was carried out. A 9.4T magnetic field Bruker Avance spectrometer with a microimaging probe was used. T_2-weighted acquisition was performed with a 16x16 μm^2 in-plane resolution and 0.2 mm slice thickness (STK), number of averages (NS) =128, matrix size (MTX)=512x512 and Field Of View (FOV) =0.8x0.8 mm^2, Repetition Time (TR) =1000 ms.

To evaluate T_2, T_2-weighted images, using the same aforementioned parameters but choosing MTX=256X256, were obtained with echo times (TEs)= 6, 12, 18, 24 ,30 ms. For T_1 evaluation, saturation recovery acquisitions at different TR: 300, 450, 700, 1000, 2000 ms were carried out (STK=0.2 mm, NS=64, MTX=256x256, FOV =0.8x0.8 mm^2, in plane resolution 32x32 μm^2, TE=3ms).

2.2. Global Diffusion, and relaxation times

To evaluate global water diffusion coefficient D, a Pulse gradient stimulated echo (PGSTE) [4] was used with TR=3s, NS=64, 32 gradient strength values (g) from 2.5 to 121 Gauss/cm, gradient pulse duration δ=3ms, diffusion time Δ=80ms.

An inversion recovery with TR=6s and 30 inversion times from t=1ms to t=4000ms was used to obtain detailed T_1 values, whereas a Carr Purcell Meiboom Gill (CPMG) with TEs=n*0.5ms (n=1 to 128) and TR=6s was used for extracting T_2 values belongings to different water pools [5,6].

2.3. Processing NMR imaging and NMR data

Three different regions of interest (ROIs) (see Figure 1) were chosen in T_2 and T_1 weighted images being interested to investigate wood features related to a different amount of water. Specifically, T_1 and T_2 were evaluated in each late-wood (dark pixels), early-wood (light pixels) ROI and in the whole sample. From imaging data, the recovery of the equilibrium magnetization (M0) along the direction of the static field was fitted to a single exponential function of the form:

$$M(TR) = M0[1-\exp(-TR/T_1)] \quad (1)$$

so, obtaining the corresponding relaxation time T_1. To obtain T_2, the signal intensity versus TE was fitted using the function:

$$M(TE)=M0*\exp(-TE/T_2) +c \quad (2)$$

On the other hand, considering the global signal (no-imaging modality) T_1 and T_2 were obtained by using bi-exponential fitting functions [5,6]:

$$M(t) = M0_1[1-2*\exp(-t/T_{11})] + M0_2[1-2*\exp(-t/T_{12})] \quad (3)$$

and

$$M(TE)=M0_1*\exp(-TE/T_{21}) +M0_2*\exp(-TE/T_{22}) \quad (4)$$

respectively.

Finally, diffusion coefficients D_1, D_2 and D_3 characterizing the diffusion of bulk water, restricted water and cellulose hydration water, were obtained by fitting PGSTE data with the function:

$$S(g)= \Psi_1*[\exp(-(\gamma g\delta)^2(\Delta-\delta/3)D_1]+ \Psi_2*[\exp(-(\gamma g\delta)^2(\Delta-\delta/3)D_2]+ \Psi_3*[\exp(-(\gamma g\delta)^2(\Delta-\delta/3)D_3] \quad (5)$$

where Ψ_1, Ψ_2 and Ψ_3 are proportional to the water molecules contained in each pool.

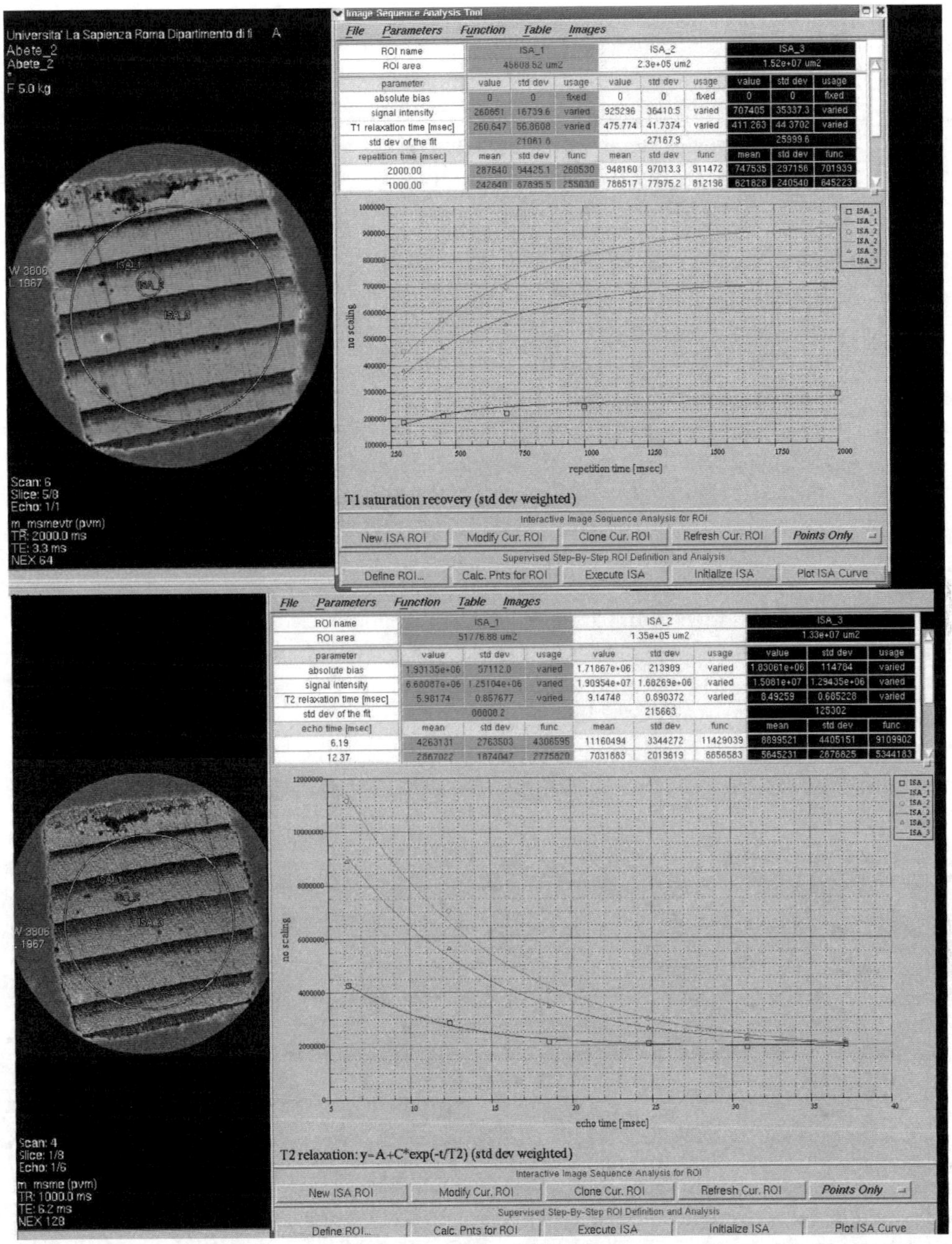

Figure 1. *Abies alba* cross-section T_1-weighted (upper image) and T_2-weighted (lower image) together with T_1 data recovery and T_2 signal decay used to extract T_1 and T_2 values, respectively. Data analysis was performed in dark and lighter pixels regions using Equations (1) and (2).

2.4. Acquisition FTIR spectra

FTIR measurements were carried out on silver fir samples with a size of about 5x5 mm^2. FTIR spectra were collected, in the 300-3000 cm^{-1} range, by means of a Perkin-ELMER spectrometer (Spectrum 100 model), using an ATR (Attenuated total reflection) equipment, which provides automatic recognition of top-plate crystal material and number of reflections. This advanced configuration incorporates features designed to minimize or even remove the sources of ordinate errors associated with standard FTIR.

3. Results and discussion

Figure 1 (upper image) displays T_1-weighted image and the signal intensity as a function of the repetition time (TR), in the two different ROIs (dark and light pixels regions) corresponding to latewood and earlywood and in an extended ROI including a mean area of the sample. From T_1 imaging data, we find T_1=(476±42ms) and T_1=(261±57ms) for the light and dark region, respectively. The value in the extended ROI gives rise to the average T_1=(411±44ms). Figure 1 (lower image) displays T_2-weighted image and the signal intensity as a function of the echo time (TE), in the two different ROIs (dark and light pixels regions) corresponding to latewood and earlywood. Note that, from T_2 imaging data, we find T_2=(9.2±0.9ms) and T_2=(6.0±0.2ms) for the white and black region, respectively. Moreover, in the mean ROI, we obtain T_2=(8.5±0.7ms). However, since the value of 6 ms coincides with the lowest TE, we have performed a more detailed investigation (see the details of CPMG pulse sequence reported in the Material and Method section) on the global sample and, using Equation 4, we obtain T_{21}=(17.0±0.8ms) and T_{22}= (2.7±0.2ms) in agreement with the results reported in references [9]. Conversely, the quantification of the longitudinal relaxation times, performed using Equation 3, gives us a single value of T_1 equal to (440±54)ms. All the obtained values reflect the state of water in the considered regions and, therefore, characterize the system under study in its specific conditions. Known the wood structure alterations induced by water, different actions could be taken to preserve it.

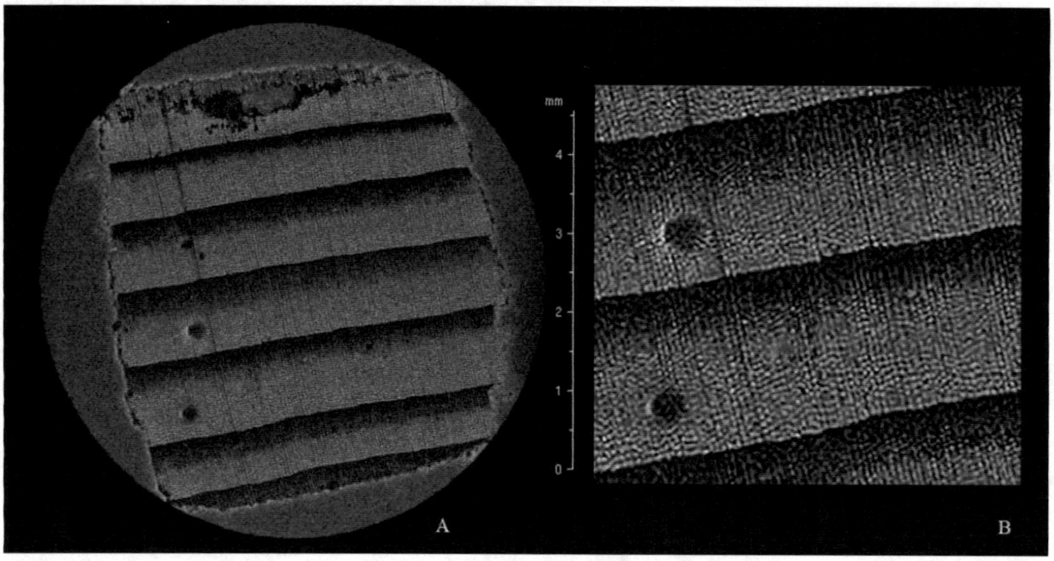

Figure 2. *Abies alba* cross-section T_2-weighted micro-MRI obtained at 9.4T magnetic field. In-plane resolution was 32x32 μm^2.

In Figure 2 micro-MRI cross-section images of the sample, *Abies alba,* is shown. The softwood structure, typical of coniferous wood, is visible. In particular, early and latewood are clearly distinguishable: latewood and earlywood correspond to dark and light pixels, respectively, while the transition between them is abrupt. Moreover, few resin canals with a diameter of about 0.25 mm are also visible. Generally, there is no presence of resin canals in silver fir; thus, we think that they refer to traumatic resin canals [7]. All these observations are in agreement with guidelines from dichotomous keys, commonly used by wood experts as reference for the microscopic identification of the wood species [8].

Figure 3 shows the signal decay, obtained using a PGSTE sequence, as a function of the gradient strength g and the best fitting function.

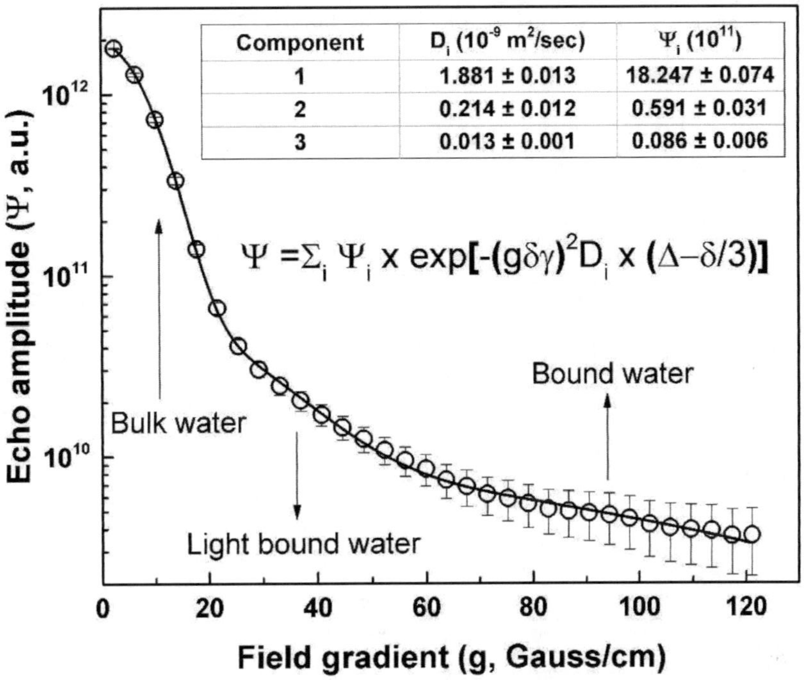

Figure 3. The decay of the echo amplitude, evaluated by means of PGSTE pulse sequence, as a function of the gradient strength g (γ= 42.58 MHz/T is the proton gyromagnetic ratio). Three exponential contributions have been identified and ascribed to bulk water, light bound water between fibres and bound water inside fibres (or cellulose hydration water).

Three different contributions, belonging to three different mobile water molecules, have been discriminated. In particular, the biggest value of the diffusion coefficient D_1 = (1.881+-0.013)*10^{-9} m^2/s belongs to bulk water (we remember that the sample is immersed in liquid water), that is the water inside vessels. D_2 =(2.140+-0.120)*10^{-10} m^2/s belongs to water at the fibres interface of the wood sample and inside the cells [10]. Finally, D_3 = (1.30+-0.10)*10^{-11} m^2/s belongs to tightly bound water molecules, strongly confined (pore on the order of nm) or bonded to hydroxyl groups of cellulose and lignin and that also remain on drying [11-12]. The normalized water population Ψ associated with estimated diffusion coefficients are: 96.4%, 3.1% and 0.5% of the total magnetization. Therefore, the 96.4% of the total observable water protons in sample belongs to bulk water, i.e. water in vessels, the 3.1% belongs to intracellular water and the 0.5% to water inside silver fir, as hydration water of cellulose and lignin.

In terms of degradation, wood pore size distribution is a key parameter in water accumulation. As shown above, diffusion NMR techniques may provide an indication on separate wood-bound water, restricted light bound water, and confined water in vessels (bulk water) fractions by measuring diffusion NMR parameters together with relaxation times T_1, T_2 estimation.

Complementary information about the water bonding properties in relation to wood structure is obtained carrying out FTIR measurements. Figure 4 and Table 1 report the recorded FTIR spectrum and the corresponding assignments, respectively. We have performed a Gaussian deconvolution with six components to reproduce the main features in the 2000-3500 cm^{-1} spectral range, detailed in Table 1. From left to right, the sub-bands are ascribed to: 1) the scissor/rocking vibration of water, 2) CO_2 stretching mode, 3) CH stretching of cellulose, 4) CH_2 stretching of cellulose and hemicellulose 5) free OH in cellulose and 6) free OH in lignin, respectively. Really the two contributions, centred at about 3200 and 3385 cm^{-1}, are also ascribed to the OH stretching bands of bound and mobile water, respectively.

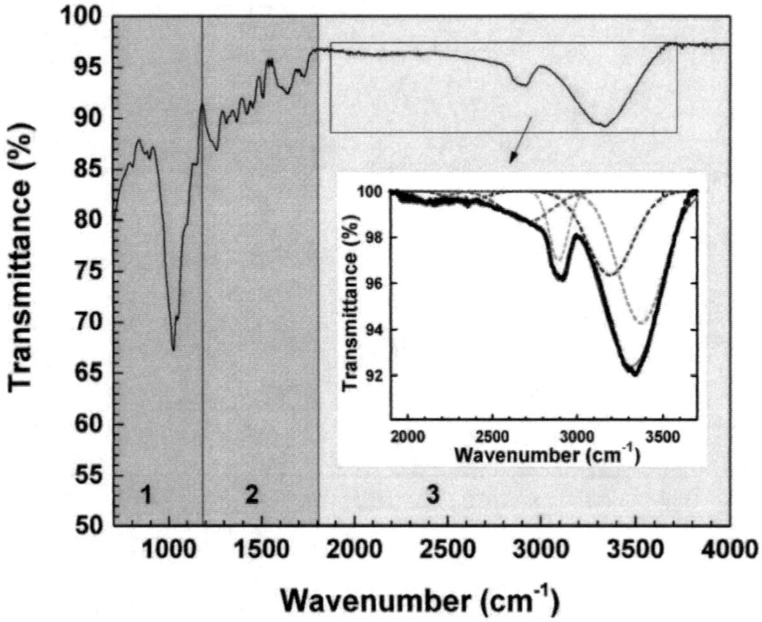

Figure 4. FTIR spectrum of the *Abies alba* wood. In the inset, the deconvolution of the profile observed in the 2000-3500 cm^{-1} range.

From the ratio between the analytic area of these sub-bands, we estimate the relative percentage of mobile water with respect to bound water, which results of about 61%. We emphasize that, when we talk about bound water (by means of hydrogen bonds) in NMR analyses, we refer to the hydration water exchanged with a fast rate on the hydration surface. Given the "exchange" times, it is not plausible, by the NMR technique, to observe very strongly bound water. Therefore, to obtain this complementary information, FTIR technique was used, which provides local information on wood chemistry and structure at the level of wood cells, alterations of the composition and interactions between wood polymers within the lignocellulose complex induced by water filtration.

Hence, FTIR has been used to determine the gradient of the water absorbed (following the hydroxyl groups OH) between the surface and the inner part of the *Abies alba* wood. In detail, the ratio between the OH contribution at 3400 cm^{-1} and the lignin C=C aromatic contribution centred at around 1610 cm^{-1} gives information about the penetration degree of OH groups from the surface than inside the wood. Generally, the ancient or artificially aged woods show a decrease of OH content from the surface to the inner part, as expected since the ageing of the surface favours the creation of sites where OH groups are fixed. In our case, this ratio is about 2.2, a value which agrees with those found in ancient woods, at a depth of about 8 mm from the surface [13].

Table 1. The main FTIR bands of the *Abies alba* wood and their assignments.

Wavenumber (cm^{-1})	Assignments
3358-3372	Free OH in lignin; free water
3195	Free OH in cellulose; bound water
2935	CH stretching of cellulose and hemicelluloses
2901	Asymmetric CH_2 stretch
2862	CH_2 stretching of cellulose and hemicelluloses
2679	CH stretching
2350	CO_2
2100	Scissoring and rocking vibrations of water
1770	C=O stretching in conjugated ketones
1735	C=O stretching of carboxylic acid in hemicelluloses ester group or in unconjugated ketones
1698	C=O vibration in the carboxylic group in resin acid
1660	C=O stretching in cellulose
1640	H-O-H angle vibration of adsorbed water (peak assigned to bound water)
1600	C=C stretching of the aromatic ring (S)
1500	C=C stretching of the aromatic ring (G)
1275	Characteristic peak of lignin
1240	C-O of acetyl in pectin or hemicelluloses
1160	anti-symmetrical deformation of the C-O-C band
1150	Symmetric C-O-C and asymmetric stretch in cellulose in hemicellulose
1112	Aromatic C-H out-of-plane deformation C=O stretch
1087	C-O; C-C; -C-H stretch in alcohol in lignin
1056	C-O stretching of secondary alcohols
1024	C-O stretching in primary alcohols in cellulose
990	C-O stretching in cellulose
964	C-H out-of-plane deformation in lignin
939	Aromatic C-H out-of-plane deformations
895	C-H deformation of beta-glycosidic linkages in cellulose
871	C-H deformation of hemicellulose
860	C-H out of the plane in position 2, 5, and 6 of guaiacyl units

On the overall, the combination of NMR and FTIR analyses allows determining wood species by the identification of the microscopic characteristics on NMR images and, at the same time, achieving information about microstructure and preservation status through relaxometry measurements and FTIR spectra. Clear information about dynamics of water or other substances inside samples, such as archaeological waterlogged woods, can be obtained. Moreover, in the case of conservation treatments, it could be a useful tool to choose the most suitable consolidant product or to evaluate its penetration. Finally, we stress that both the considered techniques are non-invasive and non-destructive, which are the main important features for cultural heritage studies.

4. Conclusion

In this work, we have used NMR and FTIR experimental techniques in order to characterize the physical-chemical conditions of silver fir, with particular emphasis on the water contribution that is the principal responsible for degradation effects. The complementary use of NMR and FTIR techniques has allowed to perform a systematic study of the principal wood morphological/structural modifications (by NMR) associated with humification processes, such as changes of aromatic C-H and C-O vibration modes, presence of phenol groups as principal substituents and a reduction in oxygen-containing functional groups, principally carboxylic or carbonylic groups (by FTIR). Then, the synergic use of these techniques, probing the system energies on different scales, is an innovative approach, able to characterize the system under study in a complete way with the aim to prevent degradation effects and to improve its preservation over time. In particular, the amount and mobility of water with respect to those of lignin and cellulose well reflect the physical state of the wood system and can serve as a

IOP Conf. Series: Materials Science and Engineering **777** (2020) 012004 doi:10.1088/1757-899X/777/1/012004

benchmark for its degradation conditions.

References

[1] www.ivalsa.cnr.it/en/research/physical-and-mechanical-characterization-of-wood-and-wood-products.html

[2] Ringman R, Beck G and Pilgård A 2019 *Forests* **10(6)** 522

[3] Bao M, Huang X, Zhang Y, Yu W and Yu Y 2016 *J. Wood Sci.* **62(5)** 441-451

[4] Stejskal E O and Tanner J E 1965 *J. Chem. Phys.* **42** 288-292

[5] Pirazzoli I, Alesiani M, Capuani S, Maraviglia B, Giorgi R, Ridi F and Baglioni P 2005 *Magn Reson Imaging* **23 (2)** 277-284

[6] Alesiani M, Capuani S, Giorgi R, Maraviglia B, Pirazzoli I, Ridi F and Baglioni P 2004 *J. Phys. Chem. B* **108 (15)** 4869-4874

[7] Franceschi V R, Krokene P, Christiansen E and Krekling T 2005 *New Phytol* **167 (2)** 353–375

[8] www.woodanatomy.ch

[9] Gezici-Koc Ö, Erich S J F, Huinink H P, Van der Ven L G J and Adan O C G 2017 *Cellulose* **24** 535–553

[10] Topgaard D and Söderman O 2002 *Cellulose* **9** 139–147

[11] Choong E T and Skaar C 1972 *Wood Fiber Sci* **2** 80-86

[12] Ben Dhib K, Elaieb M T, Azzouz S and Elcafsi A 2016 *Mater. Environ. Sci.* **7 (7)** 2561-2571

[13] Vartanian E, Barres O and Roque C 2015 *Spectrochim Acta A* **136** 1255-1259

Cyclodextrins in topical gel formulation as photoprotective system for Nabumetone

M A Occhiuzzi, F Grande*, M De Luca, G Ragno, A Garofalo and G Ioele*

Department of Pharmacy, Health and Nutritional Sciences, University of Calabria, 87036 Arcavacata di Rende (CS) - Italy

* E-mail: giuseppina.ioele@unical.it; fedora.grande@unical.it

Abstract. Photostability studies applied on aqueous solutions of Nabumetone, an anti-inflammatory drug used in the treatment of osteoarthritis and rheumatoid arthritis, have confirmed the sensitivity to light of this compound, revealing the 6-methoxy-naphthalene-aldehyde as the main photoproduct. In this work, the stability of NA-cyclodextrin (CD) complexes was investigated in gel formulation potentially suitable as sustained-release systems. The photodegradation experiments were realized under stressed conditions according to the ICH rules and monitored by spectrophotometry. The spectral data were processed by Multivariate Curve Resolution (MCR), able to estimate spectra and concentration profiles of the components involved in the kinetic process. NA entrapped in cyclodextrin and formulated in solution and gel preparations were exposed to an irradiance power of 350 W/m^2. Encapsulation percentage of the drug in several cyclodextrins was measured, recording an increase of the water solubility in the order hpβCD>mβCD>βCD. No significant photoprotection of NA was measured in aqueous solution. On the contrary, the gel containing the hpβCD-complex showed relevant stability. The photoprotective ability of this formulation was further increased by adding ascorbic acid 2%, still detecting 90% of the starting concentration after 90 min of light exposure.

1. Introduction

The safety of the drugs, during production, storage, distribution and use by patients, is affected by exposure to several causes of degradation, including temperature and light [1]. In particular, the exposure to light could affect the physicochemical properties of either a pure compound or a pharmaceutical formulation. A great amount of liquid or in topical gel preparations, liquid or in gel, are classified as sensitive to light in the U.S. and European Pharmacopeia [2], consequently, the development of advanced photoprotective systems is still demanding. Furthermore, the use of appropriate protective containers or packaging may also be involved in the photostabilization strategies of the drugs [3]. When the pharmacokinetic behavior of a drug is not favorable and the absorption through the skin take too long, an active compound should be also protected after application, allowing a prolonged exposure to sunlight. In these cases, several approaches of encapsulation into supramolecular systems have been proposed, very often supported by the addition of antioxidants and solar filters [4-7]. In the pharmaceutical field, cyclodextrin (CD) complexes represent the most used encapsulation matrices to realize photoprotective carriers [8]. These cyclic oligosaccharides can incorporate lipophilic drugs into the hydrophobic cavities by non-covalent

IOP Conf. Series: Materials Science and Engineering **777** (2020) 012005 doi:10.1088/1757-899X/777/1/012005

complexation [9]. Moreover, the variety of available CDs, both natives and modified, allows the incorporation of drugs with different size.

In this work, the light stability of Nabumetone (4-(6-methoxy-2-naphthyl)butan-2-one) (NA) was investigated in solution and gel formulations. Photoprotection of NA was performed after entrapping the drug into CD matrices. According to the ICH rules, the photodegradation tests were made in an appropriate standard irradiation chamber on drug solutions and CD complexes in solution and gel formulations.

The photodegradation profiles were monitored by spectrophotometry and the spectral data were processed by Multivariate Curve Resolution (MCR). This chemometric procedure was chosen because particularly suitable for following transformation kinetic processes as it is able to estimate spectra and concentration profiles of the components involved [10-13].

2. Materials and methods

2.1. Chemicals, Instruments and software

NA, ascorbic acid (AA), propylene glycol, microcrystalline cellulose, βCD, mβCD, and hpβCD were purchased from Sigma-Aldrich (Milan, Italy). Ethanol and methanol were from J.T. Baker (Holland).
UV spectra were recorded by using a Perkin-Elmer Lambda 40P Spectrophotometer by setting the following instrumental conditions: λ range 200–450 nm, scan rate 1 nm/s; time response 1 s; spectral band 1 nm. Spectral acquisition and elaboration were made by using the dedicate software UV WinLab® (Perkin-Elmer, Waltham, MA).

A light cabinet Suntest CPS+ (Heraeus, Milan, Italy) equipped with a Xenon lamp was used to perform the photodegradation experiments. The ID65 standard filter was set to simulate sunlight in a spectral range between 300 and 800 nm.

Multivariate analysis was performed by the software Matlab® computer environment (Mathwork Inc., version 7).

2.2 Sample preparation

NA standard solutions were prepared in ethanol at the concentration of 5.0 μg/ml in consideration of the low solubility of the drug in water. CD-complexes were prepared by dissolving an excess of NA (25 mg) in 10 ml of cyclodextrin solution (10mM) and 10.0 ml of Britton-Robinson buffer pH 6.57 (0.04 M phosphoric acid; 0.04 M acetic acid; 0.04 M boric acid; 0.2 M NaOH) under stirring for 20 h at 37° C. Four series of samples (4 x 5) were prepared by using βCD, mβCD, and hpβCD. All the samples were stored for 4 days at 4° C and then filtered through a 0.45 μm membrane. Spectrophotometric measurements were performed after dilution 1:10 of the samples in ethanol. For the solubility tests, the drug content was added to each cyclodextrin solution with concentrations of 1, 5, 7.5 and 10 mM in 10 ml buffer.

Gel formulation (20 g) was prepared according to the European Pharmacopoeia [14]. NA 0.20 g (1% w/w) was emulsified in propylene glycol 2 g under continuous stirring for 15 minutes. 0.60 g of microcrystalline cellulose (gelling agent) and 17.2 g of water were then added and the final emulsion was stirred for 50 min getting a homogeneous white gel.
The gel formulation containing the complex CD-drug was prepared by adding 17.2 g of the CD complex solution to propylene glycol and microcrystalline cellulose.

A last sample was prepared by adding AA to the CD-drug complex before the preparation of the gel, in such a way to obtain a percentage concentration of 2%.

2.3. Photodegradation test

The photodegradation tests were made on all the prepared liquid and gel formulations under the following conditions: irradiation power 450 W/m2, corresponding to 27 kJ/m^2 min, temperature 25 °C. The samples were analysed just after preparation (t= 0 min) and at the several exposure times (10-30-

50-70-100-130-150-210-270-300 min) by UV spectrophotometry. All laboratory experiments were carried out in a dark room to minimize drug photodegradation.

The drug content in gel along the photodegradation experiments was measured by MCR applied to the UV data of the methanol extracts. At this aim, gel 0.5 g was uniformly stratified on a glass plate to form a layer thickness of 0.25 mm and then exposed to forced irradiation. After each irradiation dose, the glass plate was sonicated in acetonitrile 25 mL for 10 min at room temperature. 10 ml of the obtained suspension were centrifuged at 5000 rpm for 10 min and the supernatant was analysed after 1:10 dilution with methanol.

3. Results and discussion

3.1. Photodegradation of NA solutions

An ethanol solution of NA 5.0 μg/ml was subdued to forced photodegradation, under the standard conditions above reported. The spectra, depicted in figure 1, were recorded just after the preparation and at several exposure times up to five hours.

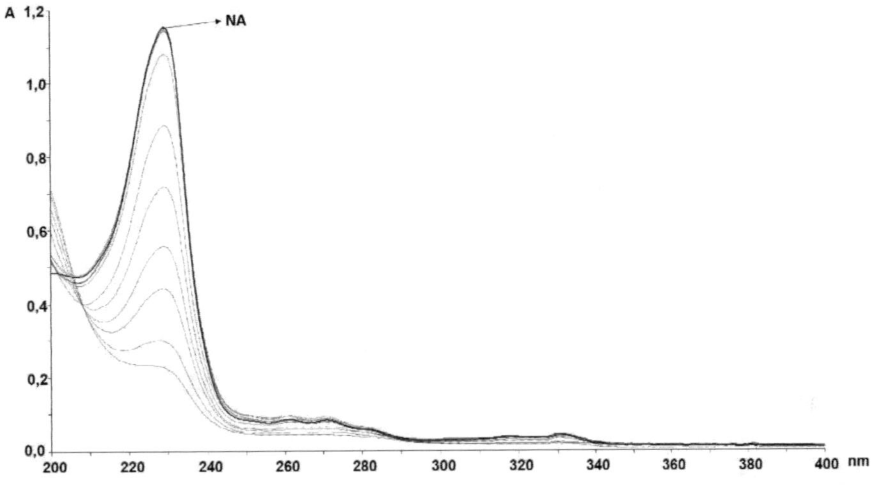

Figure 1. Absorbance spectra recorded for the ethanol solution of NA at several exposure times.

The spectral data, as an average of five experiments collected during the photodegradation tests, were used to construct the data matrix to be analysed by MCR-ALS. The results reported in figure 2 showed the photodegradation profiles of NA and its photoproducts (A) and the respective absorbance spectra (B).

IOP Conf. Series: Materials Science and Engineering **777** (2020) 012005 doi:10.1088/1757-899X/777/1/012005

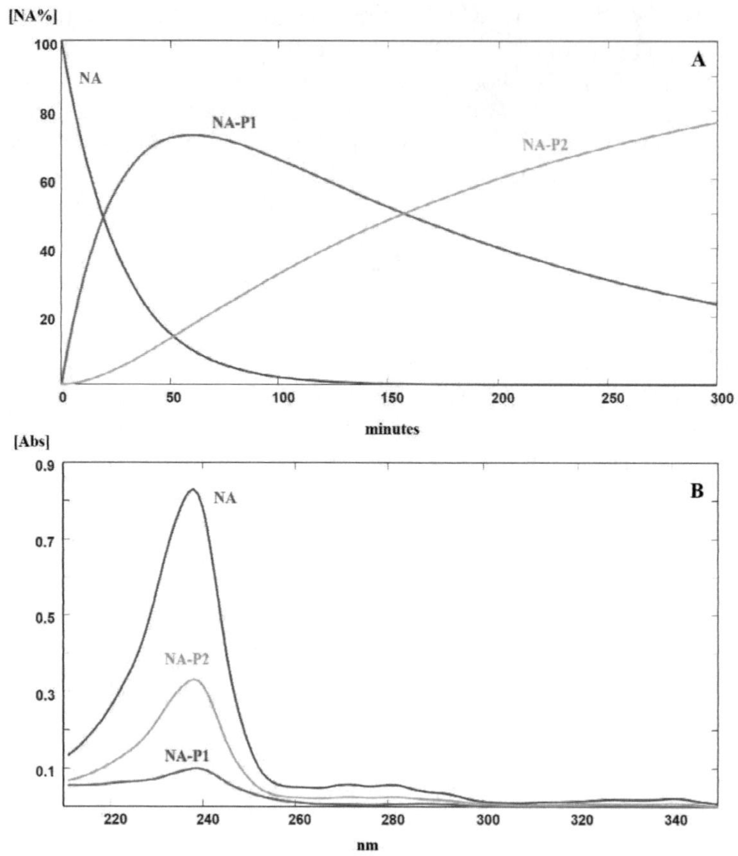

Figure 2. Photodegradation profiles of NA and its photoproducts (A) and their absorbance spectra (B).

In accordance with the results reported in the literature, data elaboration confirmed the formation of one major photoproduct (NA-P1) and traces of another by-product (NA-P2). [15] reported the photo-oxidation of the side chain to 6-methoxy-2-naphthaldehyde, as a major product, in butanol solution and the formation of the (4-(6-methoxy-2-naphthyl)-3-buten-2-one) (figure 3).

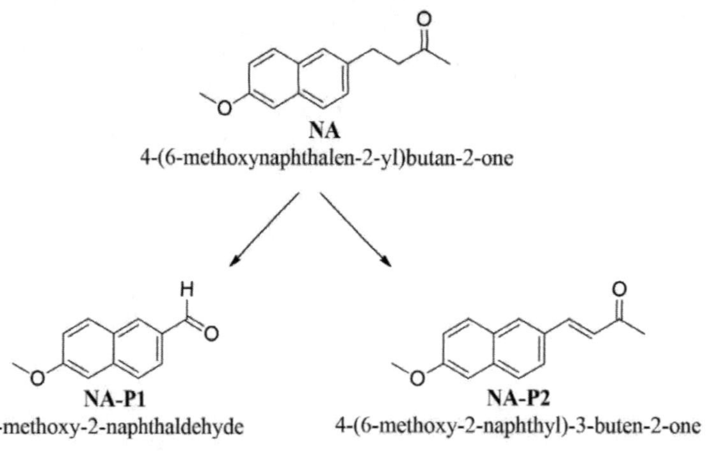

Figure 3. Chemical Structures of **NA** and its photoproducts **NA-P1** and **NA-P2**.

IOP Conf. Series: Materials Science and Engineering **777** (2020) 012005 doi:10.1088/1757-899X/777/1/012005

This photodegradation process followed the first-order kinetics. In the used solvent, the process seemed to be more efficient than in water [16].

In our experiment, the full degradation of the drug was observed after about 30 min. The degradation process proceeded via first-order kinetics, described by the equation:

$$\ln [\%NA] = - k \cdot t + 4.6$$

where %NA was the percentage of residual drug, k the photodegradation rate constant, t the time (min), and 4.6 the logarithm of the starting concentration (100%).

The parameter $t_{0.1}$ (time to cause 10% degradation) was chosen as a criterion to compare the degradation behaviour of the tested samples. This parameter is conventionally adopted because a drug could no longer be used when its purity falls below 90%. The value of k was 0.0482 and $t_{0.1}$ resulted in being 2.08 min, as reported in Table 1.

Table 1. Degradation kinetic parameters calculated for NA in different solutions.

Samples	**Entrapment efficiency (%)**	**K x 10^{-3}**	**$t_{0.1}$ (min)**	**R^2**
NA-free	-	48.2	2.08	0.999
NA-βCD	14	23.9	4.18	0.965
NA-mβCD	15	20.6	4.85	0.970
NA-hpβCD	88	4.20	23.8	0.987
NA-gel	-	23.4	4.27	0.977
NA- hpβCD gel	-	1.99	50.25	0.999
NA- hpβCD AA gel	-	1.11	90.09	0.922

In order to minimize light degradation, the stability of NA was investigated by entrapping the drug in cyclodextrin matrices. The CD systems can also enhance the solubility of the drugs in water. In fact, the influence of several cyclodextrins on the solubility of NA has been tested, as described by [17].

A set of NA aqueous solutions was prepared as above described by using βCD, mβCD, and hpβCD, respectively. The most effective cyclodextrin in increasing the solubility of NA resulted in being hpβCD. The incorporation percentage was measured by spectrophotometry as 88, 15 and 14% for hpβCD, mβCD and βCD, respectively. The best performance of hpβCD can be explained by a better fitting of the drug molecule in the cavity of this CD. All the prepared complexes were exposed to light. Figure 4 shows the photodegradation profiles of NA in the different CDs and Table 1 lists the kinetics parameters calculated by MCR.

All the used CDs increased the stability to light of the NA. In particular, the hpβCD-complex showed a $t_{0.1}$ value of 23.9 min, better than that for the simple solution but still unsatisfactory for the goals we had set.

IOP Conf. Series: Materials Science and Engineering **777** (2020) 012005 doi:10.1088/1757-899X/777/1/012005

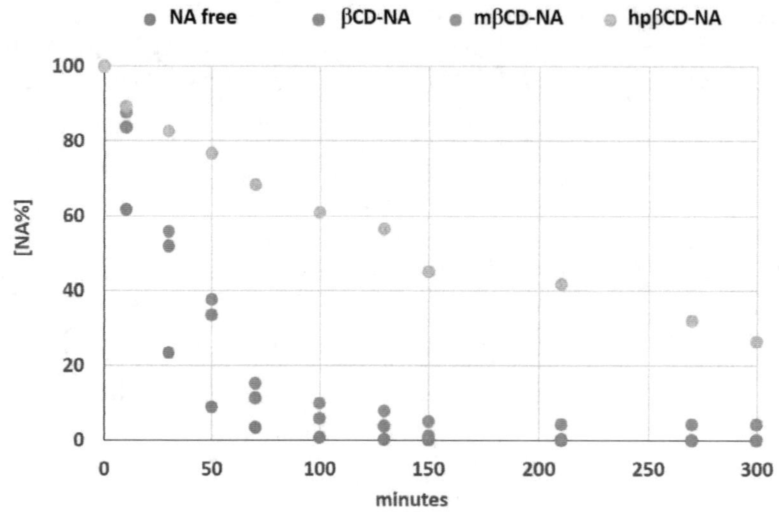

Figure 4. Photodegradation profiles of NA in the different CDs compared with the ethanol solution.

3.2. Photodegradation of NA in gel

Photodegradation tests were then applied to the above-described formulations prepared in gel. Firstly, gel formulation was made with 1% of the pure drug and exposed to light. The data obtained from the photodegradation experiments were processed by MCR method and the kinetics parameters are listed in Table 1. Also, in this formulation, the drug resulted very sensitive to light showing a $t_{0.1}$ value of 4.27 min.

A promising result was obtained when the complex hpβCD-NA was emulsified in gel. In this case, a clear decrease in NA degradation was measured, with a very successful $t_{0.1}$ value of 50.25 min. The good performance of this formulation in terms of photostabilization could be attributed to a dual-action of the CD complex: real physical protection of the entrapped drug by means of a molecular shield aided by the increase of the drug solubility in this matrix.

The light-stable formulation was optimized by adding the antioxidant ascorbic acid (2%) to the complex hpβCD-NA in gel, showing a very high increase of the stability with a $t_{0.1}$ value of 90.09 min. The photodegradation profile of this formulation hpβCD-NA-AA followed the first-order kinetics and was compared in figure 5 to that of NA 1%, as gel control, and hpβCD-NA gel. Table 1 summarizes the degradation rate constants and the values of $t_{0.1}$ for all the studied matrices.

IOP Conf. Series: Materials Science and Engineering **777** (2020) 012005 doi:10.1088/1757-899X/777/1/012005

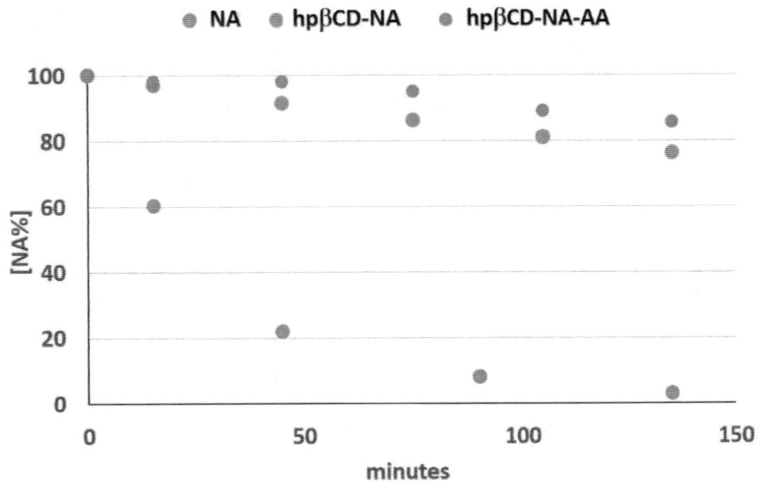

Figure 5. Photodegradation profiles of NA, hpβCD-NA complex and hpβCD-NA-AA formulated in gel.

4. Conclusions

In this work, the anti-inflammatory drug Nabumetone, either in solution or in gel formulations, was demonstrated to undergo photodegradation. The study was performed by adopting photostability tests defined by international rules. In particular, under an irradiance power of 350 W m^{-2}, corresponding to 21 kJ m^{-2} min and at a constant temperature of 25° C, the drug resulted degraded of 10% in only 2.08 and 4.27 min, in ethanol and gel, respectively. The design of photoprotective pharmaceutical matrices for topical application is particularly important due to the greater probability of light exposure that can cause a lower bioavailability of the drug and, at the same time, an increase of the risk of formation of toxic photoproducts. The entrapping of the drug into cyclodextrins and the addition of an antioxidant agent was the adopted approach to reduce drug photodegradation. The addition of ascorbic acid to the hpβCD-NA gel complex gave good results with a considerable increase of the drug photostability, reaching a very satisfactory value of 90.09 min for $t_{0.1}$. The proposed system which has been shown to be particularly effective in reducing the photodegradation of the drug could be considered a valuable starting point for the development of innovative pharmaceutical formulations for topical use of Nabumetone.

Acknowledgments
This research was supported by grants from M.U.R.S.T. (Italy).

References

[1] Tonnesen H H 2001 Formulation and stability testing of photolabile drugs *Int.J. Pharm.* **225 (1-2)** 1-14
[2] Baertschi S W, Clapham D, Foti C, Kleinman M H, Kristensen S, Reed R A, Templeton A C and Tonnesen H H 2015 Implications of In-Use Photostability: Proposed Guidance for Photostability Testing and Labeling to Support the Administration of Photosensitive Pharmaceutical Products *J Pharm. Sci.* Part 2: Topical Drug Product **104(9)** 2688-2701
[3] De Luca M, Ioele G, Spatari C and Ragno G 2016 Photostabilization studies of antihypertensive 1,4-dihydropyridines using polymeric containers *Int.J. Pharm.* **505(1-2)** 376-382

[4] Ragno G, Risoli A, Ioele G, Cione E and De Luca M 2006 Photostabilization of 1,4-dihydropyridine antihypertensives by incorporation into β-cyclodextrin and liposomes *J. Nanosci. Nanotechnol.* **6** 2979-2985
[5] Ragno G, Cione E, Garofalo A, Genchi G, Ioele G, Risoli A and Spagnoletta A 2003 Design and monitoring of photostability systems for amlodipine dosage forms *Int.J. Pharm.* **265(1-2)** 125-132
[6] Ioele G, De Luca M and Ragno G 2014 Photostability of barnidipine in combined cyclodextrin-in-liposome matrices *Future Med. Chem.* **6(1)** 35-43
[7] Ioele G, De Luca M, Garofalo A and Ragno G 2017 Photosensitive drugs: a review on their photoprotection by liposomes and cyclodextrins *Drug deliv.* **24 (sup1)** 33-44
[8] Coelho L, Almeida I F, Sousa Lobo J M and Sousa E S J P 2018 Photostabilization strategies of photosensitive drugs *Int. J. Pharm.* **541(1-2)** 19-25
[9] Scalia S, Casolari A., Iaconinoto A and Simeoni S 2002 Comparative studies of the influence of cyclodextrins on the stability of the sunscreen agent, 2-ethylhexyl-p-methoxycinnamate *J. Pharmaceut. Biomed.* **30(4)** 1181-1189
[10] Ragno G, Vetuschi C, Risoli A and Ioele G 2003 Application of a classical least-squares regression method to the assay of 1,4-dihydropyridine antihypertensives and their photoproducts *Talanta* **59 (2)** 375-782
[11] De Juan A, Rutan S C, Maeder M and Tauler R 2009 *Comp. Chem.* Vol. 2 (Oxford, U.K.: Elsevier) pp 207-558
[12] De Luca M, Mas S, Ioele G, Oliverio F, Ragno G and Tauler R 2010 Kinetic studies of nitrofurazone photodegradation by multivariate curve resolution applied to UV-spectral data *Int.J. Pharm.* **386(1-2)** 99-107
[13] De Luca M, Tauler R, Ioele G and Ragno G 2013 Study of photodegradation kinetics of melatonin by multivariate curve resolution (MCR) with estimation of feasible band boundaries *Drug Test. Anal.* **5(2)** 96-102
[14] *Council of Europe,* 2010 *European pharmacopoeia* (Strasbourg, France: 9 th Edition)
[15] Valero M 2004 Photodegradation of Nabumetone in n-butanol solutions *J. Photochem. Photobiol. A: Chemistry* **163(1-2)** 159-164
[16] Valero M and Costa S M B 2003 Photodegradation of Nabumetone in aqueous solutions *J. Photochem. Photobiol. A: Chemistry* **157(1, 21)** 93-101
[17] Higuchi T and Connors K 1985 Phase solubility techniques *Adv. Anal. Chem. Instr.* **4** 127-212

IOP Conf. Series: Materials Science and Engineering **777** (2020) 012006 doi:10.1088/1757-899X/777/1/012006

New perspectives for the green economy in Sicily

F Parisi[1,*], G Sabatino[1], G Marcianò[1], A Mottese[1], G Nania[2], F Leonetti[2], S Sfameni[3], M Di Bella[4], P Mazzoleni[5], G Barone[5], A Tripodo[1], D Drommi[6] and S Magazù[1]

[1] Department of Mathematical and Computational Sciences, Physical Science and Earth Science, University of Messina, Viale Ferdinando Stagno d'Alcontres 31, 98166 Messina, Italy.
[2] Edil Ponti SCARL, Contrada Giardina Km 98 300 snc, Gela (Caltanissetta), Italy.
[3] Department of Engineering, University of Messina, C.da Di Dio, Vill. S. Agata, 98166 Messina, Italy.
[4] Istituto Nazionale di Geofisica e Vulcanologia (INGV), Sezione di Palermo, Via Ugo la Malfa 153, 90146 Palermo, Italy.
[5] Department of Biological, Geological and Environmental Sciences, University of Catania, Corso Italia 57, 95129 Catania, Italy.
[6] Department of ChiBioFarAm, University of Messina, Viale Ferdinando Stagno d'Alcontres 31, 98166 Messina, Italy.

* E-mail: ing.parisifrancesco@gmail.com

Abstract. Geopolymers are synthetic materials, which attract increasing interest because they represent a supplementary cementitious material as an alternative to Portland cement. Geopolymers are considered as environmentally friendly materials, due to the low processing temperature and the absence of CO_2 gas emissions. These ecological features, linked to their technical properties, such as high strength, high acid resistance, and/or high-temperature resistance, make them very innovative technological materials. In addition, geopolymers show good performance if realized by the utilization of secondary raw materials (industrial wastes like fly ash or slag), thus improving strong interest in such technology from countries with growing industrialization. Here, in order to reduce global impacts and to stimulate the Sicilian green economy, we provide the evaluation of the Life Cycle Assessment (LCA) through the employ of local raw materials to produce geopolymers in Sicily. To reach this aim, geopolymers have been produced in collaboration with a Sicilian cement industry, through the use of local raw materials (furnace blast slag from Sicilian steelworks and Calabrian kaolinite) and the construction of a pilot plant. The obtained results show, for different scenarios, a considerable reduction of both CO_2 emissions and energy consumption, but also a general improvement of the environmental indicators.

1. Introduction

"Geopolymers" are synthetic materials obtained from alkaline activation of natural raw materials (e.g. clay as kaolin/metakaolin) or secondary raw materials (industrial wastes like fly ash or blast furnace slag) used to make binder systems as an alternative to Portland cement, besides restoration applications in the field of cultural heritage [1].The production process of Portland cement is well acknowledged as having a significant impact on the environment, for the high amount of CO_2 emissions (5-8% of global anthropic emissions) and the substantial energy consumption (mainly from fossil fuels) [2].

This study aims to analyze environmental and energetic impacts for the production of 1 kg of Geopolymers based concrete in two different formulations in a cement industry (Edil Ponti SCARL) in Sicily, Italy. The goal is to provide local raw materials (a furnace blast slag from Sicilian steelworks and a Calabrian kaolinite clay) to reduce global impacts and to stimulate the Sicilian green economy.

The most usable and complete technique to evaluate the potential impact on the environment during the Geopolymers production is named as Life Cycle Assessment (LCA) [3]. LCAs can help avoid a narrow outlook on environmental concerns by: 1) definition of goals; 2) compiling an inventory of relevant energy and material inputs and environmental releases; 3) evaluating the potential impacts associated with identified inputs and releases; 4) interpreting the results to help make a more informed decision.

The method used for LCA in this study is "from cradle to gate" which means that are evaluated environmental impacts associated with all the stages of the product life, from raw material extraction and distribution, through materials processing and manufacturing to the industry gate.

2. Local raw materials and analytical techniques

2.1. Sicilian furnace blast slag

To boost the Sicilian green economy, it is vital to choose local raw materials supplying. The furnace blast slag comes from a Sicilian steelworks (Acciaierie di Sicilia Spa, Catania) through a furnace blast (bow model). The geopolymerization reaction of the slag is highlighted by the different pattern resulting from the XRD characterizations (Figure 1) of the material before and after alkaline activation.

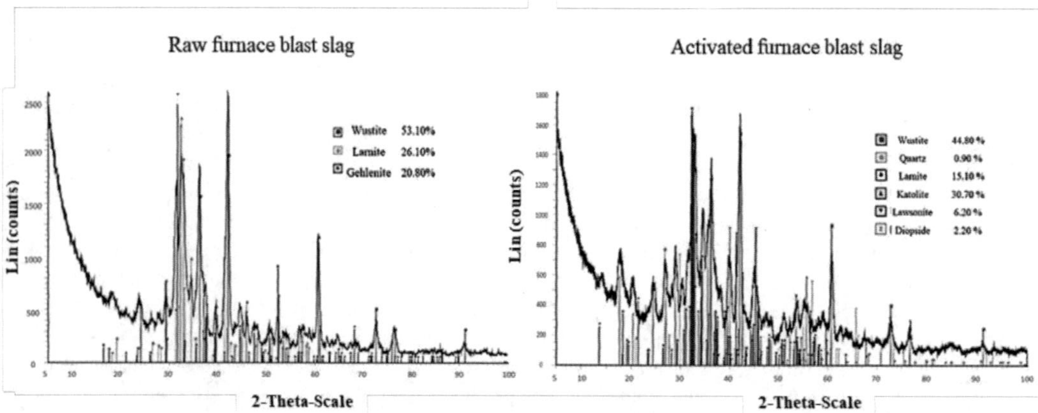

Figure 1. XRD patterns of the raw furnace blast slag (left) and of activated furnace blast slag (right).

2.2. Calabrian kaolinite

In Calabria there are several sites where kaolin is extracted. Two quarries in which mining activities are still active are in Acri (Cs), named "Colle Costantino" and "Colle Costantino Ampliata".The site under study falls within the area of Longobucco, where mainly terrigenous lithotypes are present. The succession is represented by the formation of Torrente Duno (conglomerates, sands and clays), and at the bottom the Hercynian crystalline basement. In the past, some studies have been conducted [4] focused on 18 samples cataloged as follows: the sediments of the Torrente Duno formation (LB1 – LB8 samples), the sediments of the Fosso Petrone formation (sample LB9 – LB11) and Fiume Trionto (sample LB12 – LB18). Table 1 shows major elements and Chemical Index of Alteration (CIA) ratio of Calabrian kaolinite samples: LB1 – LB8 samples have content of SiO_2, Al_2O_3, TiO_2, Fe_2O_3, and K_2Osimilar to PAAS (Post-Archean Australian Shale) [5] and a lower MnO, MgO, CaO,Na_2O, and P_2O_5 content; the sediments of the Fosso Petrone formation (sample LB9 – LB11) and Fiume Trionto (sample LB12 – LB18) result rich of CaO and poor of SiO_2, Al_2O_3, TiO_2, Fe_2O_3, MnO, Na_2O, K_2O andP_2O_5 related to the PAAS.

Table 1. XRF data of major elements (wt %) and Chemical Index of Alteration (CIA) ratio of the analyzed calabrian kaolinite.

Formations	Torrente Duno							
Sample	**LB1**	**LB2**	**LB3**	**LB4**	**LB5**	**LB6**	**LB7**	**LB8**
SiO_2	55.16	71.27	65.92	59.38	58.81	59.23	63.32	71.18
TiO_2	0.90	1.10	1.03	1.30	1.35	1.20	0.93	0.95
Al_2O_8	22.08	16.15	19.20	20.05	20.73	19.49	19.52	13.94
Fe_2O_8	6.04	2.35	2.49	6.79	1.01	4.03	5.01	3.35
MnO	0.01	0.01	0.02	0.02	0.02	0.01	0.01	0.01
MgO	1.60	1.11	1.51	1.43	1.48	1.00	1.28	1.44
CaO	0.30	0.14	0.51	0.18	0.19	1.06	0.22	0.81
Na_2O	0.16	0.14	0.31	0.17	0.18	0.32	0.16	0.47
K_2O	4.72	3.51	3.70	4.44	4.66	3.72	4.44	3.53
P_2O_5	0.03	0.02	0.02	0.04	0.04	0.03	0.02	0.01
L.O.I.	9.01	4.19	5.59	6.22	5.65	9.92	5.08	4.31
C.I.A.	73.40	73.30	73.50	73.00	72.70	71.70	73.30	65.70

Formations	Fosso Petrone			Fiume Trionto						
Sample	**LB9**	**LB10**	**LB11**	**LB12**	**LB13**	**LB14**	**LB15**	**LB16**	**LB17**	**LB18**
SiO_2	54.15	64.83	64.32	57.13	47.91	46.98	60.99	39.60	37.32	30.45
TiO_2	0.75	0.61	0.56	0.69	0.68	0.68	0.53	0.42	0.41	0.40
Al_2O_8	16.85	10.14	9.19	12.23	13.15	13.62	9.33	8.48	7.19	8.90
Fe_2O_8	3.83	3.52	3.06	3.50	3.33	3.24	2.82	2.56	2.72	2.78
MnO	0.02	0.03	0.03	0.01	0.02	0.02	0.03	0.03	0.04	0.04
MgO	2.26	2.05	1.98	2.19	2.53	2.47	1.99	1.72	1.70	2.12
CaO	5.03	6.23	7.77	8.62	12.95	12.71	9.79	21.47	24.23	25.00
Na_2O	0.30	0.42	0.43	0.35	0.35	0.34	0.36	0.27	0.35	0.19
K_2O	3.70	2.74	2.45	2.76	2.76	2.80	2.59	2.01	1.49	2.15
P_2O_5	0.06	0.08	0.07	0.04	0.04	0.05	0.08	0.08	0.09	0.09
L.O.I.	13.04	9.36	10.14	16.27	16.27	17.09	11.50	23.36	24.47	27.89
C.I.A.	-	-	-	-	-	-	-	-	-	-

3. Life Cycle Analysis

3.1. Production process description

The preparation of geopolymeric concrete is made up of three principal steps: raw materials supplying, distribution to the industry and preparation process.

To realize the final product, the first step is to grind the raw materials through an electronic mill Elite MGS Italy (1 kW) model with four independent grinding unit of 1800 cc for 30 min. After grinding, the raw materials have the eligible granulometry to be mixed by a cement mixer Syntesi 190 IMER (1 kW) for 30 min. Then, the mixture is placed manually on formworks and put in an electric oven (3 kW) at 80°C for 72 h. Table 2 shows the compositions of the two samples analyzed for 1 kg of final product(declared unit of LCA):

Table 2. Compositions of geopolymeric concretes for LCA.

Furnace blast slag concrete 1 kg		**Calabrian Clay concrete 1 kg**	
Furnace blast slag	225 g	Calabrian clay	225 g
Sand	770 g	Sand	770 g
Sodium silicate (37%) (dry)	0,72 g	Sodium silicate (37%) (dry)	9,25 g
Sodium hydroxide 9 M (dry)	16,20 g	Sodium hydroxide 12,5 M (dry)	5 g
Water	40 ml	Water	100 ml

3.2. LCI Life Cycle Inventory

Through the Life Cycle Inventory are evaluated input and output data for all the stages of the product's life. Table 3 shows the global impacts for 1 kg of raw material supplying (S) and distribution (D):

Table 3. CO_2 emissions and Energy demand for raw materials supplying (S) and transport (D).

Raw materials (1 kg)	**Emission (Kg CO_2/kg)**		**Energy demand (MJ/kg)**	
	S	D [6]	S	D [6]
Furnace blast slag [7]	0,551	0,0085 (100 km)	18,7	1687,5
Calabrian clay [8]	0,424	0,036 (420 km)	4	6930
Sodium silicate (37%) (dry) [9]	0,424	0,65(1350 km)	13,14	4050
Sodium hydroxide (dry) [10]	1,100	0,024 (310 km)	8,35	5759
Sand (extraction) [8]	0,424	-	4	-

The production process, for both samples, is influenced by thermal curing at low temperatures (80°C for 72 h) as results from the Table 4:

Table 4. CO_2 emissions and energy demand for production process.

Production process (1 kg)	**Emission (Kg CO_2/kg)**	**Energy demand (MJ/kg)**
Grinding (30 min)	0,025	0,180
Mixing (30 min)	0,025	0,180
Thermal curing (80°C,72 h)	0,480	3,450
TOT.	0,530	3,810

3.3. Calculation of geopolymer concrete impacts

To evaluate environmental and energetic impacts of the final product (Global Warming Potential – GWP, and Cumulative Energy Demand - CED), the input data for each raw material are calculated according to the relative presence in the compositions of the two samples:

Table 5. GWP and CED indicators for the two geopolymeric concretes.

LCA for 1 kg of furnace blast slag concrete			
		GWP ($kgCO_{2eq}$)	**CED (MJ)**
Raw materials supplying		0,15	4,45
Distribution		0,04	865,60
Production process		0,53	3,81
	TOT.	**0,72**	**873,86**

LCA for 1 kg of calabrian clay concrete			
		GWP ($kgCO_{2eq}$)	**CED (MJ)**
Raw materials supplying		0,06	0,16
Distribution		0,02	658,30
Production process		0,53	3,81
	TOT.	**0,61**	**662,27**

4. Results and discussion

In this study are analyzed the environmental impacts of the production of two geopolymeric concretes through the Life Cycle Analysis.

In terms of CO_2emissions, for both samples, the thermal curing, in the production process, is the principal source. Thanks to the low temperature (80 °C), GWP of both geopolymeric concretes is 40%-50% lower than Portland concrete manufacturing (1 $kgCO_2eq/kg$), on average. Both samples have shown the possibility of room temperature curing so it is possible to break down the value of GWP and CED indicators.

The furnace blast slag is an ideal raw material for the production of a geopolymeric binder by alkaline activation. In Sicily, it is easily available and its production is steady throughout the year. The goal is to make it a secondary raw material otherwise it would be destined for landfill disposal.

The result of LCA shows that the cumulative energy demand of the entire life cycle depends mostly on distribution. For distances over 500 km the impacts are remarkable. To minimize the environmental impacts and to boost the sicilian green economy, it is vital to choose local raw materials (overall for chemicals supplying). Further studies are needed to extend laboratory results to an industrial upscaling.

5. Conclusions

The aim of this work is to value the production of two geopolymeric concretes in Sicily (Italy). Local raw materials like calabrian kaolin or sicilian furnace blast slags are used to realize two different types of concrete. Analytic characterizations show the eligible properties of raw materials for the geopolymerization process.

Geopolymeric concretes have mechanical performances comparable to ordinary Portland concrete and are used to make binder systems in the constructions or for restoration applications in the field of cultural heritage.

A Life Cycle Assessment (LCA) "from cradle to gate" is drafted for the realization of each samples and two global indicators are evaluated (GWP, CED). Results show that each sample produces a GWP indicator 40% lower than Portland concrete manufacturing (1 kg CO_2eq/kg) on average. Geopolymer production represents a smart and eco-friendly alternative to Portland cement, which allow the production of a green material and also improving the sicilian local economy.

Acknowledgments

The research is supported by the AGM for CuHe project. PNR 2015-2020. Area di Specializzazione "CULTURAL HERITAGE". CUP E66C18000380005. CNR and Noxorsokem Group S.r.l. are gratefully acknowledged for S. Sfameni Industrial PhD funding.

References

[1] Provis J L and van Deventer J S J 2009 *Geopolymers Structures, Processing, Properties and Industrial Applications* (Woodhead Publishing Limited and CRC Press LLC) pp 2-6
[2] Provis J L and van Deventer J S J 2009 *Geopolymers Structures, Processing, Properties and Industrial Applications* (Woodhead Publishing Limited and CRC Press LLC) pp 379-380
[3] Provis J L and van Deventer J S J 2009 *Geopolymers Structures, Processing, Properties and Industrial Applications* (Woodhead Publishing Limited and CRC Press LLC) pp 194-209
[4] Perri F, Cirrincione R, Critelli S, Mazzoleni P and Pappalardo A 2008 Clay Mineral Assemblages and Sandstone Compositions of the Mesozoic Longobucco Group, Northeastern Calabria: Implications for Burial History and Diagenetic Evolution, *Int Geol Rev* **50** 1116–1131
[5] Taylor S R and McLennan S M 1985 The continental crust: Its composition and evolution, United States
[6] Bernetti A, Contaldi M and Sestili P 2017 ISPRA Istituto superiore per la protezione e la ricerca ambientale *Transport Rep 2017*
[7] Buttiens K, Leroy J, Negro P, Thomas J S, Edwards K and De Lassat Y 2016 The Carbon Cost of Slag Production in the Blast Furnace *J Sustain Metall* **2** 65-66
[8] Lavagna M, Mondini D and Paleari M 2011 Murature ad alte prestazioni. Valutazioni termiche, acustiche, ambientali ed economiche di soluzioni di involucro in laterizio (Maggioli Editore) pp 89-90
[9] Fawer M, Concannon M, and Rieber W 1999 Life Cycle Inventories for the Production of Sodium Silicates *Int. J. LCA* **4** 210-212
[10] Bousted I 2005 Eco-profiles of the European Plastic Industry: Sodium Hydroxide, A report for Plastic Europe Association, pp 1-18.

IOP Conf. Series: Materials Science and Engineering **777** (2020) 012007 doi:10.1088/1757-899X/777/1/012007

Sol-gel based coatings for the protection of cultural heritage textiles

V Trovato[1,*], G Rosace[1], C Colleoni[1], S Sfameni[2], V Migani[1] and M R Plutino[3]

[1] Department of Engineering and Applied Sciences, University of Bergamo, Viale Marconi 5, 24044 Dalmine (Bg), Italy
[2] Department of Engineering, University of Messina, C.da Di Dio, Vill. S. Agata, 98166 Messina, Italy
[3] Institute for the Study of Nanostructured Materials, ISMN–CNR, O.U. Palermo, c/o Department of ChiBioFarAm, University of Messina, Viale F. Stagno d'Alcontres 31, Vill. S. Agata, 98166 Messina (Me), Italy

* E-mail: valentina.trovato@unibg.it

Abstract. In this paper, the procedure for the development of sol-gel silica coatings for the protection of cultural heritage textiles, using a multistep approach, is described. With this aim, Tetraethoxysilane precursor was used at two different concentrations to realize coatings onto textile samples, using Dibutyltindiacetate as polycondensation catalyst for sol-gel reactions. The relationship between the presence of the catalyst, the number of coating layers deposited on cotton fabrics, and the influence of such architecture on several properties of treated textiles, were investigated. In particular, the chemical structure of thin films and their morphology were studied by infrared spectroscopy and scanning electron microscopy, respectively. The thermo-oxidative properties and the abrasion resistance of sol-gel treated cotton fabrics were studied, as well as the durability of the coatings after different washing cycles. Both silica coatings provided high adhesion onto cotton textiles, while the higher silica concentration resulted in higher thermal stability and washing fastness of treated fabrics. An enhancement in the washing fastness and abrasion resistance for the two different concentrations of precursor in the presence of the catalyst was also observed. All experimental findings demonstrated the efficiency of sol-gel based materials specifically tailored to the protection and preservation of cultural heritage textiles.

1. Introduction

Ancient and archaeological textiles are a precious historical-artistic heritage, which must be protected against the deterioration with time induced by environmental agents, such as light, temperature and humidity changes, as well as atmospheric pollutants. Generally, the materials employed in the preservation or restoration of textiles with historical and artistic value, are commercial products, with properties not specifically developed for the conservation of natural polymers [1,2]. Graft copolymerization of vinyl monomers onto linen and cotton fabrics were successful investigated due to their consolidating and protective effects [3]. Recent scientific advances led to a wide array of new coatings based on nanostructured materials tailored for conservation and protection of the cultural heritage. Among them, many scientific reports are concerning the application of such materials, synthesized by the sol-gel method, in the conservation and consolidation of stones and glasses [4].

According to this technology, it is possible to realize advanced materials with a high degree of homogeneity at molecular level and unique physical-chemical properties. Based on subsequently hydrolysis and condensation reactions, this versatile and cheap synthetic strategy, performed at low temperature, conducts to inorganic or hybrid organic-inorganic networks by starting from metal alkoxide precursors [5,6]. The thin silica films can be applied as protective coatings on materials surface, among which textile fabrics: it has been found that the colloidal silica improves the mechanical properties of the substrate allowing good optical transparency, and provides further features such as hardness, anti-corrosion, wear resistance and so on [7,8]. Indeed, the hybrid organic-inorganic sol-gel coatings are able to improve or confer several fabric performances, such as flame retardancy [9-11], UV protection [12], water repellency [13,14], as well as antimicrobial [15,16], and halochromic properties [6,17-22], among many.

Tetraethoxysilane (TEOS) is a silica inorganic alkoxide widely employed in the sol-gel technique and several studies have been conducted to study the adhesion of silica coatings on the surface of different materials such as glasses, plastics, metals, and textiles [23,24]. The employment of catalysts, both acids or bases, strongly influences the gelation process of TEOS [25] and also leads to adverse effects mainly consisting of the potential aggressiveness for treated material. In this regard, recently, Dibutyltindilaurate (DBTL) and Dibutyltindiacetate (DBTA), usually employed for stone conservation applications, have been investigated as organic-tin catalysts for polycondensation of sol-gel reactions [26,27]. In this work, Tetraethoxysilane and Dibutyltindiacetate (the sol-gel precursor and polycondensation catalyst, respectively) were used at different concentrations to realize protective coatings onto cellulose-based textile samples. With this aim, a multilayer approach consisting of the deposition of 1-to-6 consecutive TEOS layers for each used concentration (0.03 M and 0.3 M), with and without the presence of DBTA as the last layer onto fabric samples, was developed to thoroughly study the properties that can be achieved on treated samples. The chemical composition, the thermo-oxidative properties, the morphology and the abrasion resistance of the treated cotton fabrics were deeply investigated as function of both TEOS concentration and numbers of layers applied onto cotton fabrics. Moreover, the durability of the coatings after 1 and 5 washing cycles was assessed. The final goal of this research is to demonstrate the potential use of the sol-gel technique in the effective protection of cellulose-based polymer used in cultural heritage textiles, opening up a large research landscape no entirely investigated.

2. Experimental part

2.1. Materials

Scoured and bleached cotton fabrics, with a mass per unit area of 200 g/m^2, were employed in this research as a reference of fabric samples used in the preservation or restoration of textiles for cultural heritage applications. Before experiments, fabric samples underwent a washing cycle in a non-ionic detergent (2%) at 40°C for 20 minutes. After rinsing with deionized water and drying, the fabrics were conditioned under standard atmospheric pressure at 65±4% relative humidity and 20±2°C for at least 24 hours. For all performed experiments, Tetraethylorthosilicate (TEOS, $\geq$ 98%), Dibutyltindiacetate (DBTA), hydrochloric acid (HCl) and ethanol were purchased from Sigma-Aldrich and used without further purification.

2.2. Synthesis of TEOS coatings and application onto cotton fabrics

A 0.3 M TEOS solution was obtained through the hydrolysis of TEOS in deionized water in the presence of ethanol (6 mL) and HCl (0.1 M, 4 mL) under vigorous stirring for 5 h at room temperature. Then, the same solution was diluted by 10 times with water to obtain a second 0.03 M TEOS solution. The DBTA (8% v/v, on TEOS amount) was used as an ethanol solution and applied on cotton fabrics as the last layer. Each TEOS solution (0.03 M and 0.3 M) was applied onto cotton fabrics by a two-roll laboratory padder with a wet pick-up of 70%. The treated samples (CO_t and CO_T, cotton fabric treated with 0.03 M TEOS and with 0.3 M TEOS solutions, respectively) were dried (90°C for 5 min), re-coated with the same TEOS solution (to obtain CO_n, where "n" refers to 1, 3 and 6 layers), or treated with the catalyst solution (CO_nC), and finally cured (160°C for 5 min). The samples were conditioned under standard conditions for at least 24 h, before all the characterizations.

The following equation (1) was used to evaluate the total dry solid add-on (wt.%):

$$Add\text{-}on = \frac{W_T - W_{UT}}{W_{UT}} \cdot 100 \quad (1)$$

In Equation (1), W_{UT} and W_T are the dry weights of untreated and treated fabrics, respectively.

The washing fastness properties of the applied coatings onto cotton surfaces were also investigated. Treated fabrics were subjected to 1 and 5 laundering cycles according to the standard EN ISO 6330:2012, and the weight loss of silica coatings after washing cycles (WLW, wt.%) was calculated according to Equation (2):

$$WLW = \frac{Add\text{-}on_w}{Add\text{-}on_{uw}} \cdot 100 \quad (2)$$

where $Add\text{-}on_w$ and $Add\text{-}on_{uw}$ are the total dry solid add-on of washed and unwashed cotton samples, respectively. All samples were weighted three times using a Mettler balance (± 10^{-4} g), and an average value was calculated with a standard deviation lower than ±2%.

2.3. Characterization techniques

Both untreated and treated cotton fabrics were investigated through FTIR spectroscopy to assess the chemical composition of the applied silica layer. A Thermo Avatar 370 spectrometer, equipped with an attenuated total reflection (ATR) accessory and a diamond crystal as an internal reflectance element, was used for these characterizations. FTIR spectra of both untreated and treated cotton fabrics were recorded in the range from 4000 to 700 cm^{-1}, acquiring 32 scans with a resolution of 4 cm^{-1} and reported in terms of Absorbance after normalization at 1315 cm^{-1} (CH_2 wagging of cellulose).

Morphology of both untreated and treated cotton fabrics was investigated through Scanning Electron Microscopy (SEM) using a JEOL instrument (model JSM-6380LV) coupled with Energy Dispersive Spectrometer (EDS) and operating at 20 kV.

Thermo-oxidative properties of all samples were assessed through Thermogravimetric analysis (TGA). These characterizations were performed by a TA Instruments Model TGA Q500 operating under N_2 or air atmosphere in the range 30–600°C, with a heating rate of 10°C/min and 60 mL/min gas flow.

Treated cotton samples were tested according to DIN EN ISO 12947-3 (Martindale method) to assess their abrasion resistance. A pressure of abrasive heads of 9 kPa was used, and the weight loss (wt.%) of samples was calculated for four abrasion levels (namely 5,000; 10,000; 15,000; 20,000; corresponding to the number of rubs) according to Equation (3):

$$Weight\ loss = \frac{W_f - Wi}{W_i} \cdot 100 \quad (3)$$

where W_i and W_f are the weight of the sample before and after abrasion cycles, respectively. All samples were weighted five times using a Mettler balance (± 10^{-4} g), and an average value was calculated with a standard deviation lower than ±1%.

An Instron Tensile Tester model 4501 was used to investigate the tensile strength of treated samples following ASTM D5035 standard. Five specimens for warp and five for weft direction were tested. Pretension of 0.5 ± 0.1 cN tex^{-1} and a load cell of 1000 cN were used.

3. Results and discussion

3.1. Add-on of treated cotton fabrics and chemical composition of TEOS coatings

Treated cotton fabrics were studied by means of gravimetric tests as a function of different TEOS concentrations, numbers of applied layers and presence of the catalyst as last layer. Gravimetric analyses reveal the same add-on (wt.%) for CO_t independently from both the number of applied layers and presence of the catalyst (Figure 1). Contrary, the higher the number of TEOS layers onto cotton fabric the higher the add-on (wt.%) of CO_T samples, with values up to 14.5 % (Figure 1) regardless the presence of the catalyst. This behavior is due to the interaction of the last TEOS coating with a previous one according to covalent bonds. The Si-OH of the silica layer onto cotton surfaces form an inorganic network highly interconnected with fibers by weak interactions whose strength influences the adhesion of the coating onto cotton fabrics. For the first silica layer, the adhesion of the coating is influenced by the diffusion of silica particles from the fabric surface into the fabric texture, while increasing the number of layers, the influence of cotton becomes less relevant and the linear increase in the add-on (wt.%) is due to covalent bonds established between two consecutive TEOS layers.

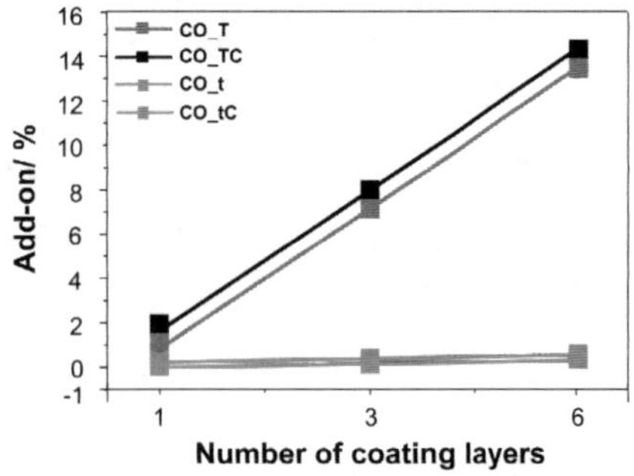

Figure 1. Add-on (wt.%) of cotton samples treated with 1, 3, and 6 layers of 0.03 M (CO_t) and 0.3 M TEOS solutions (CO_T), with and without the catalyst.

Both ATR-FTIR spectra of CO_t and CO_T are similar, and differences in the absorption intensity between untreated and treated cotton samples are more evident for CO_T samples because of the higher concentration of the deposited silica finishing. For this reason, only FTIR spectra of CO_T fabrics are reported in Figure 2. The characteristic absorption bands of cellulose in the range 3500 – 3000 cm^{-1} (O-H stretching), 2980 -2800 cm^{-1} (C-H stretching) and 1433 – 1428 cm^{-1} (C-H wagging) do not show significant changes in their intensity after the deposition of TEOS coatings [28]. FTIR peaks at 1202 cm^{-1}, 953 cm^{-1}, 796 cm^{-1} (Si-O stretching vibration, Si-OH stretching, and Si-O-Si symmetric stretching, respectively), confirm the presence of the TEOS layers onto the cotton surface [18,20]. By referring to the silica coatings, only the Si-O-Si asymmetric stretching peak (1018 cm^{-1}) [17,19] was not distinctly observed because of its overlapping by cellulose absorption bands in the range 1050 - 980 cm^{-1}. As shown by spectra in Figure 2, the presence of DBTA catalyst does not influence the FTIR profile of treated samples significantly.

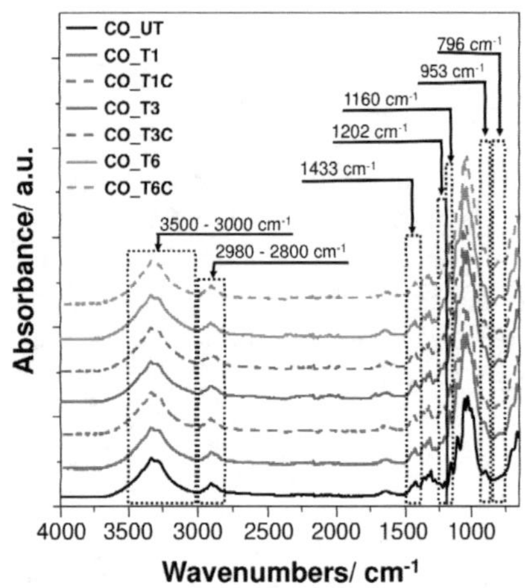

Figure 2. Comparison between ATR-FTIR spectra of CO_UT and CO_T from 1 to 6 silica layer application, with and without the catalyst (dashed and solid curves, respectively).

3.2. Washing fastness

The durability of the silica coatings onto cotton substrates was assessed in terms of washing fastness after 1 and 5 launder cycles as described in the experimental part and average WLW (wt.%) values for both CO_t and CO_T samples are reported in Table 1. Results revealed weight losses ranging from less than 1 up to 3 % in the absence of the catalyst with lower values for CO_T compared to those of CO_t samples. The presence of the catalyst reduces the weight loss of all samples below 1% thus confirming that DBTA enhances the durability of the coating by creating a compact and more durable surface onto fibers and that the silica film is firmly attached to the cotton surfaces.

Table 1. Weight loss (WLW, wt.%) for CO_t and CO_T after 1 and 5 washing cycles (1w and 5w, respectively) at different number of layers (1 to 6) with and without the catalyst.

Sample	WLW (%)		Sample	WLW (%)	
	1w	5w		1w	5w
CO_t1	3.2	3.3	CO_T1	0.9	0.9
CO_t3	1.2	1.9	CO_T3	0.6	1.1
CO_t6	1.0	1.6	CO_T6	0.7	1.4
CO_t1C	0.6	0.6	CO_T1C	0.5	0.6
CO_t3C	0.3	0.6	CO_T3C	0.5	0.7
CO_t6C	0.4	0.5	CO_T6C	0.6	0.6

3.3. *Morphological properties*

The morphology of treated cotton samples and the influence of the multi-layer architecture on it were investigated through SEM characterizations, and images of both 1 and 6 layers CO_t and CO_T samples, with and without the catalyst, are reported in Table 2. Moreover, EDS analysis was performed on treated samples and results, in terms of Si and Sn amount (wt.%), are reported also in Table 2 and referred to the presence of silica coatings and the catalyst on fabrics, respectively. SEM images confirm that the thin TEOS layer is more detectable increasing the silica layers, while the presence of DBTA does not affect the morphology of treated cotton samples significantly. EDS analysis revealed the higher Si amount for CO_T with increasing TEOS layers, and the presence of Sn in the cotton samples treated with the catalyst.

Table 2. SEM images of 1 and 6 layers CO_t and CO_T, with and without catalyst, with the corresponding Si and Sn amount (wt.%) from EDS analysis.

Sample	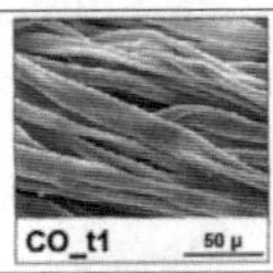		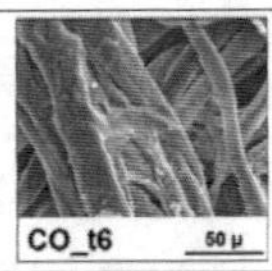	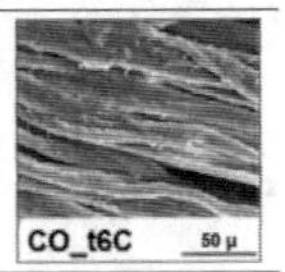
Si [%]	1.9 ± 0.2	2.0 ± 0.2	5.2 ± 0.4	5.2 ± 0.5
Sn [%]	-	0.5 ± 0.1	-	0.9 ± 0.1
Sample	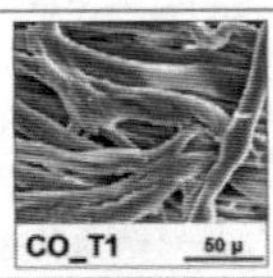	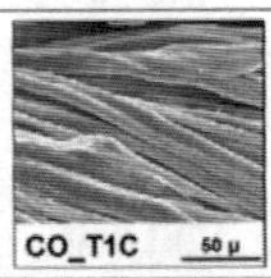	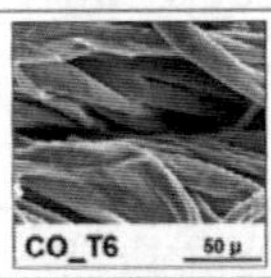	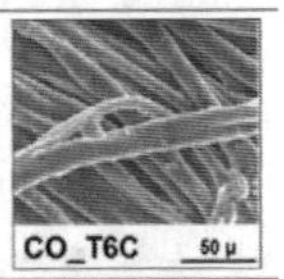
Si [%]	6.2 ± 0.6	5.2 ± 0.5	9.9 ± 0.9	8.2 ± 0.8
Sn [%]	-	1.0 ± 0.1	-	1.5 ± 0.1

3.4. *Thermo-oxidative properties*

In Table 3, the thermo-oxidative behavior of untreated and treated cotton fabrics with both employed TEOS concentrations are reported in terms of the temperature of the maximum weight loss (T_{dec}) and char formation (char residue). As shown by data reported in Table 3, the main thermo-oxidative degradation of untreated cotton fabric occurs at 336°C, to then completely burn at 600°C, as shown by the negligible char residue. The decomposition temperatures of treated samples are lower than that of untreated cotton, except for CO_t samples in absence of DBTA. The char residue percentages of CO_T samples are higher compared to CO_t thus suggesting good thermal stability of 0.3 M TEOS coating on cotton. Moreover, an increase in the char residue was observed for CO_T samples with increasing the number of layers. The presence of the catalyst seems to do not affect the decomposition temperature of these samples, while it contributes to the reduction of the char residue. An opposite effect of DBTA was observed for cotton samples treated with the lower TEOS concentration: catalyst provided a decrease in the decomposition temperature and an increase in the char residue compared to cotton samples treated without catalyst. According to these results, CO_t revealed a similar thermal degradation of untreated cotton and lower char residue than that obtained from CO_T samples. These experimental data confirm the efficacy of the higher silica concentration for the improvement of thermo-oxidative performances of cotton fabrics.

Table 3. Decomposition temperatures ($T_{dec,}$ °C) and char residue (%) at 600°C of cotton samples in air.

Sample	**T_{dec} (°C)**	**Char residue (%)**	**Sample**	**T_{dec} (°C)**	**Char residue (%)**
CO_UT	336	< 0.5	CO_UT	336	< 0.5
CO_t1	336	0.3	CO_T1	332	3.1
CO_t3	336	1.0	CO_T3	320	12.0
CO_t6	336	0.9	CO_T6	315	24.8
CO_t1C	336	2.5	CO_T1C	329	2.6
CO_t3C	332	2.5	CO_T3C	320	9.2
CO_t6C	322	5.1	CO_T6C	322	13.9

3.5. Abrasion resistance and mechanical properties of treated cotton fabrics

Martindale tests caused the broken of cotton samples treated with 0.3 M TEOS concentration, and for this reason in Figure 3, only data of Martindale tests for CO_t are reported. As shown by the graph in Figure 3, the general trend of Martindale tests consists of the increase in the weight losses (wt.%) of both untreated and treated cotton samples with increasing abrasion cycles. At the end-point of the Martindale test (20,000 cycles), the weight loss of about 14 wt.% was measured for CO_UT. The weight losses of treated samples were lower than untreated cotton fabrics thus obtaining about 11 wt.% for CO_t1 and CO_t3, and 9 wt.% for CO_t6. The presence of the last DBTA layer onto cotton surfaces led to a lowering in their weight loss compared to the same samples without the catalyst. Therefore, the best increase in abrasion resistance was noted for cotton treated with only one TEOS layer and with the catalyst (CO_t1C). This result is probably due to the better adhesion between DBTA and TEOS and between TEOS and substrate.

The influence of TEOS thin films on the tensile properties of cotton fabrics, such as strength and elongation at break, are reported in Table 4. Compared to untreated cotton, sol-gel treated textile samples show mechanical properties slightly changed. In particular, cotton fabrics finished using 0.03 M TEOS sol concentration slightly enhanced the tensile strength values, especially in the presence of three layers, in combination with the catalyst. At these conditions, a maximum weft elongation percentage is registered. No significant variations for elongation (%) performances were observed increasing the number of silica layers. This behavior is probably due to the low TEOS concentration used in the finishing that allows the fibers to be quite free to move and slide between themselves. The increase of the TEOS concentration (0.3 M), simultaneously with the number of layers, does not permit to observe improvements either in tensile strength and in elongation results. Indeed, under these conditions, both mechanical properties are generally lower than the one observed for 0.03 M sol concentration.

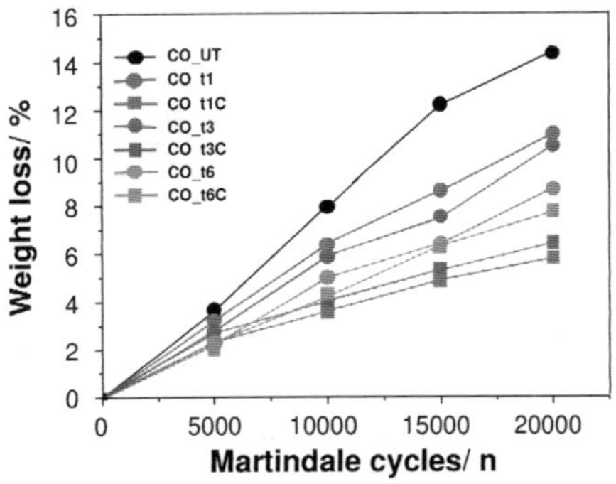

Figure 3. Weight loss (wt.%) of CO_UT and CO_t samples from 1 to 6 silica layers, with and without DBTA, as a function of Martindale cycles.

These results are in agreement with the influence of finishing on the substrate rigidity. For higher TEOS concentrations, polymerization of inorganic coating onto the textile fabrics results in a rigid layer that prevents the movement of the fibers, thus decreasing both the tensile strength and the elongation of the yarn.

Table 4. Tensile strength and elongation for untreated and treated cotton fabrics with the increasing number of layers and in the presence of the catalyst, in both warp and weft directions (standard deviation lower than ±3%). The best results are highlighted in bold.

		CO_UT	**CO_t1**	**CO_t3**	**CO_t6**	**CO_t1C**	**CO_t3C**	**CO_t6C**
Tensile strength (N)	Warp	240.3	325.2	302.4	317.9	317.2	***357.1***	282.3
	Weft	265.5	225.5	***326.6***	302.4	250.3	230.2	306.7
Elongation (%)	Warp	17.0	***17.5***	16.3	15.5	16.3	16.3	14.5
	Weft	16.0	15.3	17.8	16.3	16.7	**17.9**	16.1
		CO_UT	**CO_T1**	**CO_T3**	**CO_T6**	**CO_T1C**	**CO_T3C**	**CO_T6C**
Tensile strength (N)	Warp	240.3	207.3	190.9	150.0	200.1	185.0	150.7
	Weft	265.5	220.1	222.4	125.5	205.7	203.1	179.2
Elongation (%)	Warp	17.0	16.8	13.8	7.5	15.2	13.8	10.5
	Weft	16.0	13.9	12.2	11.0	14.0	13.9	13.5

4. Conclusions

The strategy performed for the development of TEOS thin coatings for application onto cotton fabrics according to a multi-layer architecture (1 to 6 layers) for the protection of cultural heritage textiles was described in this study. The influence of two TEOS concentrations, the role of the multilayer architecture and the effect of the DBTA catalyst as the last layer on the properties of cotton fabrics were thoroughly investigated. Treated textiles revealed good adhesion of the multi-layered silica architecture. The presence of TEOS layers onto cotton fabrics and of Si and Sn were investigated by ATR-FTIR spectroscopy and SEM-EDS characterization, respectively. The best performances in terms of washing fastness and thermal stability are shown by the cotton samples treated with the higher TEOS concentration. However, the cellulose samples treated with the lower silica concentration revealed higher abrasion resistance and mechanical properties than that showed by the cotton textiles treated with higher silica coating concentration. With increasing the number of silica layers onto cotton fabrics, both the durability and the thermal stability of final textiles increased. Moreover, durability and abrasion resistance of silica coatings are enhanced by the presence of the catalyst as the last layer, while the latter results in improved thermal stability of the coatings contained the lower TEOS concentration. Experimental findings demonstrated the efficiency of the sol-gel technique in the enhancement of specific properties of textile materials. The developed silica coatings represent reliable alternatives to existing materials thus highlighting the potential use of the sol-gel technique in the effective protection on cellulose-based polymer in cultural heritage textiles.

Acknowledgments
CNR and Noxorsokem Group S.r.l. are gratefully acknowledged for S. Sfameni Industrial PhD funding.

References

[1] Humphrey B 1986 *J. Am. Inst. Conserv.* **25** 15–29
[2] Butler C E, Millington C A and Clements D W G 1989 *Historic Textile and Paper Materials II* (Oxford: Oxford University Press) chapter 3 pp 34-53
[3] Princi E, Vicini S, Pedemonte E, Arrighi V and McEwen I J 2006 *J. Appl. Polym. Sci.* **103** 90–99
[4] Carmona N, Wittstadt K and Römich H 2009 *J. Cult. Herit.* **10** 403–409
[5] Mahltig B and Textor T 2008 Nanosol and Textiles (World Scientific Publishing Co. Pte. Ltd)
[6] Trovato V, Colleoni C, Castellano A and Plutino M R 2018 *J. Sol-Gel Sci. Techn.* **87** 27–40
[7] Plutino M R, Colleoni C, Donelli I, Freddi G, Guido E, Maschi O, Mezzi A and Rosace G 2017 *J. Colloid Interf. Sci.* **506** 504–517
[8] Rosace G, Canton R and Colleoni C 2010 *Appl. Surf. Sci.* **256** 2509–16
[9] Rosace G, Castellano A, Trovato V, Iacono G and Malucelli G 2018 *Carbohyd. Polym.* **196** 348–358
[10] Castellano A, Colleoni C, Iacono G, Mezzi A, Plutino M R, Malucelli G and Rosace G 2019 *Polym. Degrad. Stabil.* **162** 148–159
[11] Brancatelli G, Colleoni C, Massafra M R and Rosace G 2011 *Polym. Degrad. Stabil.* **96** 483–490
[12] Xing Y, Yang X and Dai J 2007 *J. Sol-Gel Sci. Techn.* **43** 187–192
[13] Colleoni C, Guido E, Migani V and Rosace G 2015 *J. Ind. Text.* **44** 815–834
[14] Colleoni C, Esposito F, Guido E, Migani V, Trovato V and Rosace G 2017 *IOP Conf. Ser. Mater. Sci. Eng.* **254** 122002
[15] Mahltig B, Fiedler D and Böttcher H 2004 *J. Sol-Gel Sci. Techn.* **32** 219–222
[16] Poli R, Colleoni C, Calvimontes A, Polášková H, Dutschk V and Rosace G 2015 *J. Sol-Gel Sci. Techn.* **74** 151–160
[17] Guido E, Colleoni C, De Clerck K, Plutino M R and Rosace G 2014 *Sensor. Actuat. B-Chem.* **203** 213–222
[18] Caldara M, Colleoni C, Guido E, Re V and Rosace G 2016 *Sensor. Actuat. B-Chem.* **222** 213–220
[19] Rosace G, Guido E, Colleoni C, Brucale M, Piperopoulos E, Milone C and Plutino M R 2017 *Sensor. Actuat. B-Chem.* **241** 85–95
[20] Plutino M R, Guido E, Colleoni C and Rosace G 2017 *Sensor. Actuat. B-Chem.* **238** 281–29
[21] Van der Schueren L, De Clerck K, Brancatelli G, Rosace G, Van Damme E and De Vos W 2012 *Sensor. Actuat. B-Chem.* **162** 27–34
[22] Caldara M, Colleoni C, Guido E, Re V and Rosace G 2012 *Sensor. Actuat. B-Chem.* **171–172** 1013–21
[23] Li F, Xing Y and Ding X 2008 *Surf. Coat. Tech.* **202** 4721–27
[24] Colleoni C, Donelli I, Freddi G, Guido E, Migani V and Rosace G 2013 *Surf. Coat. Tech.* **235** 192–203
[25] Sequeira S, Evtuguin D V, Portugal I and Esculcas AP 2007 *Mater. Sci. Eng. C.* **27** 172–179
[26] Salazar-Hernandez C, Zarraga R, Alonso S, Sugita S, Calixto S and Cervantes J 2009 *J. Sol-Gel Sci. Techn.* **49** 301
[27] Canevali C, Chiodini N, Morazzoni F, Padovani J, Paleari A, Scotti R and Spinolo G 2001 *J. Non-Cryst. Solids* **32** 293–295
[28] Chung C, Lee M and Choe E K 2004 *Carbohyd. Polym.* **58** 417-420

AUTHOR INDEX